Plant Strategies and Vegetation Processes

Plant Strategies and Vegetation Processes

J. P. GRIME

Unit of Comparative Plant Ecology (N.E.R.C.),
Department of Botany,
University of Sheffield

JOHN WILEY & SONS

Chichester · New York · Brisbane · Toronto

Library of Congress Cataloging in Publication Data:

Grime, John Philip.
Plant strategies and vegetation processes.

Bibliography: p.
Includes index.
1. Botany–Ecology. I. Title. II. Title: Vegetation processes.
QK901.G84 581.5 78-18523
ISBN 0 471 99695 5 (cloth)
ISBN 0 471 99692 0 (paper)

Typeset in VIP Baskerville by Preface Ltd, Salisbury, Wilts
and printed in Great Britain at The Pitman Press, Bath

Preface

In this book I have attempted to analyse in simple terms the processes which control the structure and composition of vegetation. The concepts and the data upon which it is based derive from three types of research which may be described, respectively, as the *correlative, direct*, and *comparative* approaches to plant ecology.

The *correlative* approach consists of the attempt to explain variation in the composition of vegetation by reference to associated environmental variation. One of the successes of this approach has been the identification of certain of the climatic and edaphic factors which determine the character of vegetation in severe environments. More recently this type of research has been extended to studies of the effects of pollutants on the distribution of vegetation types, species, and genotypes.

The *direct* approach relies upon detailed observation and recording of the establishment, longevity, and reproduction of individual plants in natural vegetation at specific sites in the field. Data which have been collected in this way allow vegetation to be interpreted as a function of events in the life-histories and population dynamics of the component plants. The strength of the direct approach is related to the opportunity which it provides to observe the process of natural selection at first hand. However, recognition, over the last decade, of the genetic and ecological fluidity of many plant populations, whilst confirming the value of intensive investigations of local populations, has also exposed the need for extreme caution in making extrapolations from such parochial studies.

The *comparative* approach involves the study under standardized experimental conditions of the germination, growth, and reproductive physiology of large numbers of species and populations of contrasted ecology. This broadly-based type of research provides a context for more localized and intensive studies, and enables the ecologist to recognize the main avenues of adaptive specialization in plants and to identify characteristics of life-history and physiology which determine fitness (or lack of fitness) in particular habitats.

In attempting a synthesis of the main results from these three fields of research it has been convenient to follow the example of many animal ecologists by focusing on 'strategies'. This form of presentation permits a more condensed style of writing and has provided several opportunities to explore ecological and evolutionary parallels between plants and animals.

During the preparation of this book, and in the preceding years of research in Sheffield and New Haven, it has been my good fortune to work with and to learn from many dedicated and talented ecologists. In particular I should like to acknowledge the guidance I have received from Dr P. E. Waggoner, Professor C. D. Pigott, the late Dr P. S. Lloyd, Dr J. G. Hodgson, Professor T. C. Hutchinson, Dr B. C. Jarvis, Dr O. L. Gilbert, and Dr R. Law. I should also like to record my thanks to many past and present colleagues and friends in the Unit of Comparative Plant Ecology, some of whom, including Professor A. R. Clapham, Dr I. H. Rorison, Dr R. Hunt, Dr A. S. Mahmoud, Dr K. Thompson, Mr M. Spray, Dr S. B. Furness, Dr M. M. Al-Mufti, Dr C. L. Sydes, Dr C. Sydes, Dr Y. D. Al-Mashhadani, Mr S. R. Band, Mr A. M. Neal, and Mr A. V. Curtis, have kindly allowed reproduction of unpublished data and photographs.

In conclusion, it is a particular pleasure to thank Mrs N. Ruttle for her immaculate typing and unfailing good humour and to record the deep debt of gratitude which I owe to my family for their constant support and encouragement throughout this venture.

University of Sheffield,
May, 1978

J. P. GRIME

Contents

PART 2 VEGETATION PROCESSES

Introduction

A large quantity of data is now available to those who seek to understand how vegetation functions and is caused to vary in composition from place to place and with the passage of time. Although it is generally recognized by plant ecologists that there is a need to deploy this wealth of information within a succinct conceptual framework, opinions differ as to how this may be achieved. The purpose of this book is to examine one approach to the problem. The method consists of the attempt to recognize the major adaptive strategies which have evolved in plants and to relate these strategies to the processes which determine the structure and species composition of vegetation.

Strategies may be defined as groupings of similar or analogous genetic characteristics which recur widely among species or populations and cause them to exhibit similarities in ecology.

The potential value of the concept of strategies as a unifying approach to ecology depends upon the extent to which it may be applied to all living organisms. Here attention will be mainly confined to autotrophic organisms, and most of the evidence which will be considered refers to vascular plants. However, the strategies described conform to basic types, several of which are evident throughout the plant and animal kingdoms.

In the attempt to define strategies, a complication arises from the need to consider different phases in the life-cycle of the same organism. Even where they experience the same habitat conditions, juvenile and mature stages within the same population may be subject to different forms of natural selection; or, alternatively, because of differences in size and function, they may respond in different ways to the same selection force.

The resultant 'uncoupling' between juvenile and mature strategies has long been recognized in animals. In many invertebrates the transition between larval and adult phases in the life-history is characterized by radical alterations in structure, physiology, and ecology, and a less dramatic but essentially similar phenomenon is evident in features such as the specialized nutrition, more strictly-defined habitat requirements, and dependence upon parental care which are usually exhibited by the offspring of vertebrate animals.

The same principle may be applied to plants, and it is clear that in order to understand the basic features of their ecology it is necessary to examine the strategies adopted during two different parts of the life-history—the established (mature) phase and the regenerative (immature) phase.

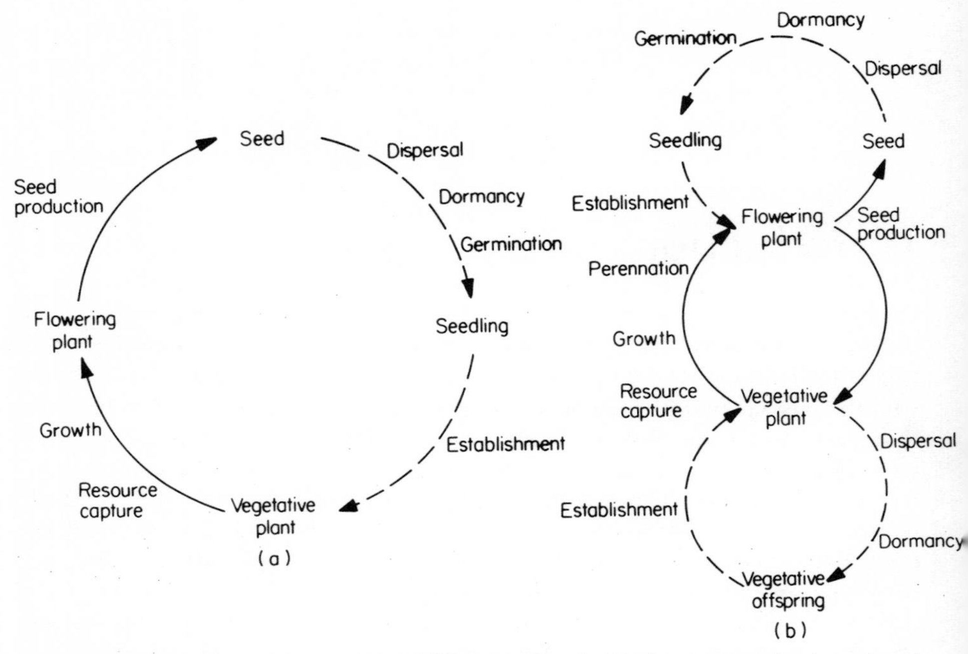

Figure 1 Schemes illustrating the activities associated with the regenerative (– – –) and established (——) phases in the life-cycles of (a) an annual flowering plant and (b) a perennial plant producing both seeds and vegetative offspring.

The scheme in Figure 1a describes the regenerative and established phases in the life-cycle of an annual flowering plant. The regenerative phase consists of a series of stages (seed release, dispersal, dormancy, germination, and seedling establishment) each of which varies in duration and in mechanism according to the species or population. The established phase is characterized by a variety of inter-related functions including the capture of resources, the maintenance, replacement, and enlargement of roots and shoots, survival of stress and damage, and the production of seeds.

The ecology of many annual plants may be interpreted as a function of two strategies which are determined, respectively, by the characteristics of the established and regenerative phases of the life-cycle. However, in certain annuals, seeds originating from the same parent may bring about quite distinct types of regeneration. Some seeds, for example, may be dispersed considerable distances to new habitats in which they germinate and establish without delay, whilst others become buried *in situ* and after a period of dormancy in the soil establish in the place originally occupied by the parent. Such diversity in the regenerative capacity of a single species may arise from genotypic variation within the seed population or may be the result of a sophisticated seed physiology. Regardless of its causation, this variety of mechanism exerts a profound effect upon the ecology of the species and must be taken into account in any strategic analysis. The importance of multiple forms of regeneration in plants is particularly clear when attention is turned to the life-cycles of perennial species, many of which reproduce both vegetatively and by seed (Figure 1b). In the majority of perennials the two types of offspring differ radically in characteristics such as size, times of production, efficiency of dispersal, and rates of mortality and for this reason they may be expected to add quite different dimensions to the ecology of a species or population.

The first two sections of this book are concerned with strategies in the established phase. Strategies in the regenerative phase are examined in Chapter 3, which also contains a review of the combinations which commonly occur between regenerative strategies and strategies in the established phase. The concluding sections explore the significance of plant strategies in relation to the processes which control the structure, dynamics, and species composition of vegetation.

PART 1

PLANT STRATEGIES

Chapter 1

Primary strategies in the established phase

INTRODUCTION

The external factors which limit the amount of living and dead plant material present in any habitat may be classified into two categories. The first, which we may describe as *stress*, consists of the phenomena which restrict photosynthetic production such as shortages of light, water, and mineral nutrients, or sub-optimal temperatures. The second, here referred to as *disturbance*, is associated with the partial or total destruction of the plant biomass and arises from the activities of herbivores, pathogens, man (trampling, mowing, and ploughing), and from phenomena such as wind-damage, frosting, droughting, soil erosion, and fire.

In the spectrum of plant habitats provided by the world's surface the intensities of both stress and disturbance vary enormously. However, when the four permutations of high and low stress with high and low disturbance are examined (Table 1) it is apparent that only three of these are viable as plant habitats. This is because, in highly disturbed habitats, the effect of continuous and severe stress is to prevent a sufficiently rapid recovery or re-establishment of the vegetation.

It is suggested that the three remaining contingencies in Table 1 have been associated with the evolution of established strategies conforming to three distinct types. These are the *competitors*, which exploit conditions of low stress and low disturbance, the *stress-tolerators* (high stress–low disturbance) and the *ruderals* (low stress–high disturbance). The three strategies are, of course, extremes of evolutionary specialization and in Chapter 2 plants adapted to habitats experiencing intermediate intensities of stress and disturbance will be examined.

First of all, however, it is necessary to consider, in turn, the three primary strategies and the evidence for their existence.

Table 1. Suggested basis for the evolution of three strategies in plants.

Intensity of disturbance	Intensity of stress	
	Low	High
Low	Competitors	Stress-tolerators
High	Ruderals	No viable strategy

COMPETITORS

A definition of competition between plants

Wherever plants grow in close proximity to each other, whether they are of the same or of different species, differences in vegetative growth, seed production, and mortality are observed. It would be a mistake, however, to attribute all such differences to the process of competition. The dangers in such an assumption are clear if it is recognized that disparities in the performance of neighbouring plants may arise from independent responses to the prevailing physical and biotic environment. It follows that if the term competition is to be useful to the analysis of vegetation mechanisms, a definition must be found which effectively distinguishes it from other processes which influence vegetation composition and species distribution.

In a most revealing paper, Milne (1961) has traced the history of biologists' attempts to define competition. He concludes that, far from achieving the necessary distinctions, a majority of authors have not sought to make them and have often used the word as a synonym for 'the struggle for existence' (Darwin 1859). It is perhaps because of the resulting confusion that Harper (1961) has proposed that, for the present at least, use of the term should be abandoned. However, it may be argued that competition as a process is too important, and as a term is too useful, to be allowed to suffer such a fate. An alternative course is to apply a strict definition and to use the term as precisely as possible.

Here, competition is defined as *the tendency of neighbouring plants to utilize the same quantum of light, ion of a mineral nutrient, molecule of water, or volume of space* (Grime 1973b). This choice of words allows competition to be defined in relation to its mechanism rather than its effects, and the risk is avoided of confusion with mechanisms operating through the direct impact of the physiochemical environment or through biotic effects such as selective predation.

According to this definition, competition refers exclusively to the capture of resources and is only part of the mechanism whereby a plant may suppress the fitness of a neighbour by modifying its environment. This dissection of the phenomena which are often 'lumped together' in the more traditional usage of the term has distinct advantages. We are able to classify the variety of mechanisms whereby plants are successful in crowded environments and, in particular, as explained in Chapter 4, we are able to analyse more satisfactorily situations in which dominance of vegetation is achieved by plants which exhibit relatively slow rates of resource capture.

Competition above and below ground

Having recognized that plants may compete for several different resources, it is necessary to consider whether any generalizations can be made with regard to the way in which the focus of competition varies according to vegetation type. In the first place, it is clear that competition for light may become a major influence upon vegetation composition only in circumstances in which the canopy is sufficiently dense for an overlap of leaves to occur. In the early stages of colonization of a fertile disturbed habitat, such as an arable field, shoots of the invading plants scarcely impinge on each other and competitive interactions, where they occur at all, are likely to be confined to those operating within the soil. As vegetation development continues and the leaf canopy closes there is opportunity for competition to occur simultaneously above and below ground.

In the case of the arable field, the relative importance of competition above and below the soil surface is a function of the maturity of the vegetation. It seems likely, however, that this relationship is characteristic only of situations in which plant colonization is allowed to proceed undisturbed in conditions of moderate to high productivity. Where vegetation is developing in a habitat of low potential productivity, such as a rock outcrop with shallow soil or where there is continuous and severe damage to the vegetation such as on a heavily trampled path, the leaf canopy remains sparse and competitive interactions will be mainly confined to the root environment.

Further analysis of the role of competition in vegetation is complicated by the long-term nature of some interactions between perennials and by the problem of distinguishing competition from other processes which affect vegetation composition. Before proceeding further, it will be helpful to identify some of the attributes which determine the competitive ability of plants.

Characteristics which determine the competitive ability of established plants

Competitive ability is a function of the area, the activity, and the distribution in space and time of the plant surfaces through which resources are absorbed and as such it depends upon a *combination* of plant characteristics. It is important to bear in mind therefore that each of the attributes considered in the review which follows is not by itself diagnostic of high or low competitive ability. As we shall see later, many of the characteristics dealt with under the following headings are capable, in other contexts, of assuming a different strategic significance.

Storage organs

Just as the competitive ability of a young seedling may be influenced by seed size (Black, 1958), so that of an established plant may be affected by the quantity of reserves stored in perennating organs. The extremely rapid expansion of the leaf canopy characteristic of many of the larger perennial herbs such as *Chamaenerion angustifolium*, *Petasites hybridus*, and *Pteridium aquilinum* (Figure 2) is the result of the mobilization of large reserves of energy and structural materials accumulated in underground storage organs (Plate 1) during the later stages of the previous growing season (Bradbury and Hofstra, 1976). An obvious competitive advantage arising from such a rapid expansion of foliage is the pre-emption of space in the leaf canopy.

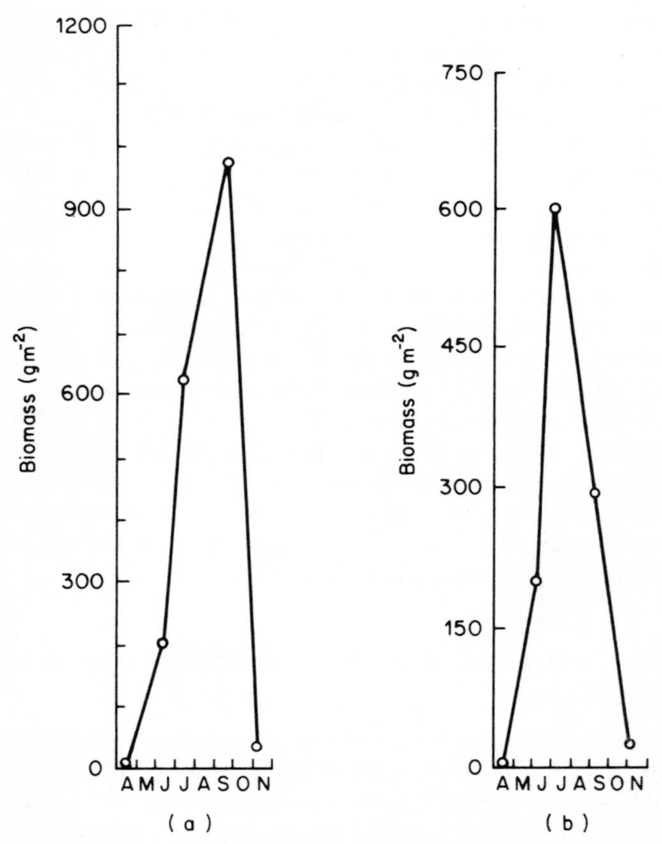

Figure 2 Seasonal change in the shoot biomass of two perennial competitors growing in fertile, relatively undisturbed vegetation in Northern England. (a) *Pteridium aquilinum*, (b) *Chamaenerion angustifolium* (Al-Mufti *et al*., 1977). (Reproduced by permission of Blackwell Scientific Publications Ltd.)

Height

Since the early studies of Boysen-Jensen (1929), it has been recognized that where perennial plants are competing for light, small differences in stature may exercise a critical effect on survival. Within closed herbaceous vegetation small differences in height are associated with large changes in the intensity, direction, and quality of radiation and the ability of a seedling or established plant to compete successfully for light may depend upon the extent to which the leaves can rapidly penetrate to superior positions in the canopy.

Height growth in the shoots of established plants is determined, first, by the supply of energy and structural materials available from storage organs or current photo-synthesis and, second, by the morphology of the shoot. Among herbaceous plants, the highest canopies occur in species in which the growing points are carried aloft at the apex of tall shoots, e.g. *Epilobium hirsutum* (Plates 2 and 3). In the majority of those species in which the growing points remain close to the ground surface there is a limited capacity for height growth. However, in certain species, e.g. *Petasites hybridus* (Plates 4 and 5), and the rhizomatous fern, *Pteridium aquilinum*, an elevated canopy is achieved through the production of a small number of massive leaves (fronds in the case of *P. aquilinum*).

Lateral spread

It is possible for a plant to reach a considerable height without capturing a major share of the resources present in its habitat. In both herbs and woody species, effective competition for light, water, mineral nutrients, and space is characteristic of species in which tall stature is allied to a growth form in which lateral expansion results in a high density of shoots and roots. Among herbaceous species the growth forms most clearly conforming to this pattern are the branching rhizomes of dicotyledonous species such as *Urtica dioica*, *Chamaenerion angustifolium*, *Solidago canadensis* and of invasive grasses such as *Holcus mollis* and *Brachypodium pinnatum* and the dense expanded tussocks of grasses such as *Arrhenatherum elatius*.

Phenology

Development of competitive attributes such as a high density of tall leafy shoots involves the production and deployment of a large quantity of photosynthate. This in turn depends upon an extended period of photosynthetic activity under climatic conditions conducive to high productivity.

In temperate climates such as that of the British Isles, productive herbs such as *Chamaenerion angustifolium*, *Petasites hybridus*, and *Pteridium aquilinum* (Figure 2) tend to attain full leaf expansion over the period June–August when daylengths, light intensities, and temperatures are favourable to high rates of photosynthesis. In some of these species the development of flowers and fruits is delayed until the late summer. This phenomenon is consistent with maximum leafiness during the period most favourable to high rates of photosynthesis, and in certain species, e.g. *Polygonum cuspidatum*, flowering may be so delayed that, at latitudes close to the northern limit of distribution, seeds often fail to ripen before the onset of winter.

Growth rate

Tall stature, extensive lateral spread, the build-up of large perennating organs, and the rapid expansion of leaf and root surface areas are all dependent upon the production annually of a large quantity of photosynthate. Estimations of the maximum relative rates of dry matter production (R_{max}) under standardized productive conditions suggest that populations of herbaceous plants from productive, relatively undisturbed, habitats tend to be among the most rapid-growing perennial species (Table 2).

Response to stress

Although high R_{max} appears to be directly advantageous in competition, its significance may extend beyond the capacity for rapid dry matter production. It may be more helpful to regard high R_{max} as one of the more easily quantified

Table 2. Estimates of the rates of dry matter production in the seedling phase in six competitive perennial herbs compared to those attained by two crop plants grown under the same conditions (Grime & Hunt 1975). Measurements were conducted over the period 2–5 weeks after germination in a controlled environment. (Temperature: 20°C day, 15°C night; Daylength: 18 h; Visible radiation: 38.0 W m^{-2}; Root medium: sand + Hewitt nutrient solution.) Rates of dry matter production are expressed as maximum (R_{max}) and mean ($\bar{R}$) relative growth rates and as a function of leaf area, i.e. as unit leaf rate (E_{max}).

Species	R_{max} week^{-1}	$\bar{R}$ week^{-1}	E_{max} g m^{-2} week^{-1}
Competitive herbaceous perennials			
Alopecurus pratensis	1.3	1.3	69
Arrhenatherum elatius	1.3	1.3	115
Chamaenerion angustifolium	1.4	1.4	80
Epilobium hirsutum	1.8	1.8	126
Phalaris arundinacea	1.2	1.2	91
Urtica dioica	2.4	2.2	89
Annual crop plants			
Lycopersicon esculentum (Tomato)	1.6	1.0	93
Hordeum vulgare (Barley)	1.6	0.9	64

expressions of a group of physiological attributes which facilitate the capture of resources. These may include, in particular, a rapid rate of response to environmental variation, especially in the extension growth of stems, petioles, and fine roots and in the expansion of leaf area. It is not difficult to foresee the extent to which rapid adjustments in morphology in response to local depletions in resources, arising during competition, will maximize the capture of mineral nutrients, water, light, or space. Short-term experiments involving shading (Grime and Jeffrey, 1965; Loach, 1970) show clearly that the extent of phenotypic adjustment, whether in leaf area or in the extension growth of stems or petioles, or in deployment of photosynthate between root and shoot, may differ considerably between species. Moreover, the rates of response are greater in potentially fast-growing species. However, it is not yet clear to what extent such differences are the result of different genetic limits to flexibility in individual leaves and roots or arise from the fact that, in many slow-growing plants, growth is intermittent and organs at a morphogenetically-responsive stage of development form a small proportion of the biomass (see pages 34–35).

Response to damage

The competitive ability of a plant may be considerably reduced where the leaf area or root surface area is subject to predation or other forms of damage. Although there have been no extensive comparative studies of plant responses to damage, field observations and the results of experiments (Milton, 1940; Mahmoud, 1973) indicate that there are various morphogenetic responses to defoliation which may be characteristic of highly competitive plants. These are particularly obvious in meadow grasses such as *Lolium perenne*, *Arrhenatherum elatius*, and *Alopecurus pratensis*, which are capable of rapid regrowth of the leaf canopy after defoliation. It would appear that the success of these plants in vegetation subjected to infrequent but severe damage by mowing (meadows and road verges) is related to their ability to respond to defoliation either by renewed growth of severed leaves or by expansion of new shoots, processes which may involve the diversion of an increased proportion of the photosynthate into shoot growth with a concurrent check in root development. It seems reasonable to suggest that such responses are specifically adapted to competition in that they cause the plant to rapidly re-establish a tall, dense leaf canopy.

Under conditions in which productive vegetation is subject to frequent mowing or grazing a slightly different phenotypic response appears to characterize strongly competitive species. In productive turf-grasses such as *Agrostis tenuis* or pasture genotypes of *Arrhenatherum elatius* (Mahmoud, Grime, and Furness, 1975), the effect of repeated defoliation is to stimulate the development of a very large number of small tillers (Figure 3) with the result that a dense and 'rapidly-repaired' leaf canopy is formed close to the ground surface.

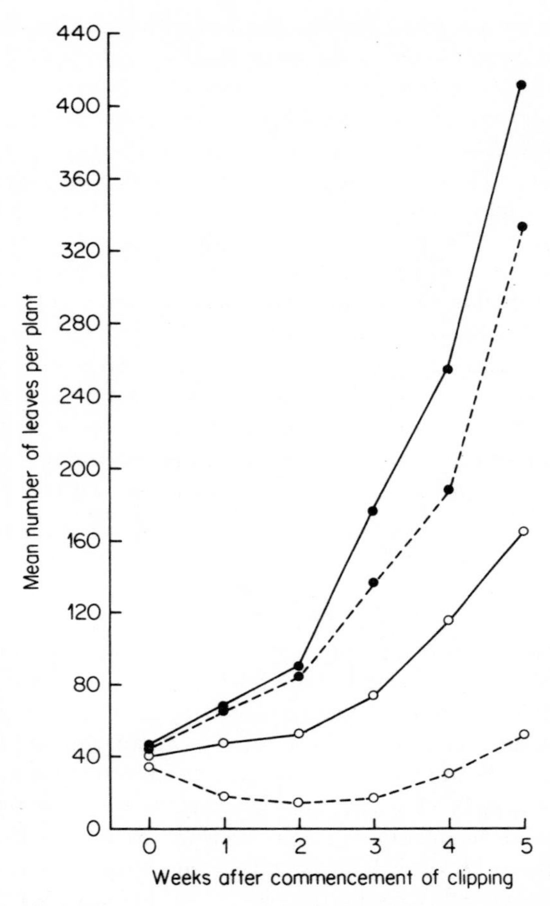

Figure 3 A comparison of leaf production under a standardized laboratory regime in unclipped (——) and frequently clipped (– – –) seedlings from a productive pasture population of *Agrostis tenuis* (•) and seedlings from a population of *Arrhenatherum elatius* (○) from tall, ungrazed grassland (Mahmoud, 1973).

Intraspecific variation in competitive ability

A fundamental problem underlying any attempt to analyse the impact of competition in vegetation arises from the proposition, scarcely disputed among ecologists, that the competitive ability of a plant species varies according to the conditions in which it is growing.

In the first place, it is clear that plant characteristics which affect the competitive ability of the species may be subject to genetic variation. The literature provides numerous examples of intra-specific variation in competitive attributes (e.g. Clausen, Keck, and Hiesey, 1940; Bradshaw, 1959; Cook, Lefèbvre, and McNeilly, 1972; Gadgil and Solbrig, 1972; Mahmoud, Grime, and Furness, 1975).

A second source of variation in competitive ability arises from the fact that environments differ in the extent to which they allow the competitive characteristics of a plant to be expressed. It is clear that the development of a dense canopy of leaves or a large absorptive surface on the root system may be restricted by forms of stress and damage. An example of modification of the competitive ability of a species by environment is illustrated in Figure 4 which compares the seasonal pattern of shoot production by bracken (*Pteridium aquilinum*) in two contrasted habitats. The results show that, at an unshaded productive, site, the vegetation consisted of a vigorous single-species stand of *P. aquilinum* in which the density of living shoot material reached a summer maximum of approximately 1000 g m^{-2}. Although the second site, situated on an acidic woodland soil, contained a high density of fronds, the dry weight of living shoot material of *P. aquilinum* at the summer maximum was restricted (presumably by mineral nutrient stress and shading) to a value corresponding to 4% of that achieved at the productive site.

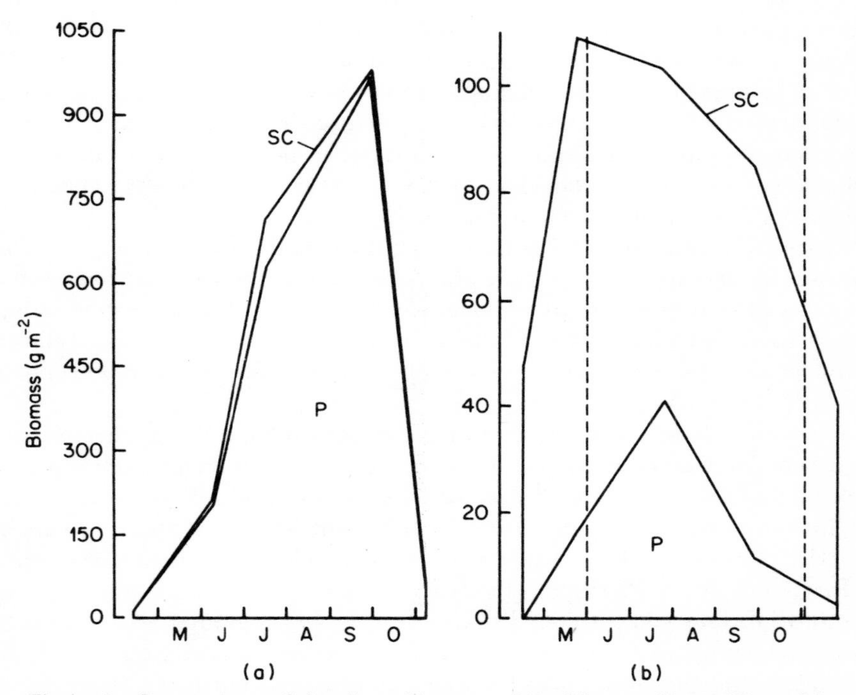

Figure 4 Comparison of the shoot biomass of *Pteridium aquilinum* (P) and its contribution to the standing crop (SC) in two contrasted habitats (Northern England, 1975): (a) in an unshaded situation on a deep brown earth soil, (b) in an oakwood on an incipient podzol (Al-Mufti *et al.*, 1977). Limits of the period of maximum shading by the tree canopy in (b) are indicated by the broken lines.

Further evidence of the modifying effect of environment and management upon competitive ability is available from a number of field experiments in which fertilizer additions have been made to the nutrient-deficient meadows (e.g. Brenchley and Warington, 1958), pastures (Smith, Elston, and Bunting, 1971), and sand-dunes (Willis, 1963). A characteristic of the responses to fertilizer applications described in these investigations is the rapid expansion of grasses such as *Arrhenatherum elatius, Holcus lanatus, Agrostis stolonifera,* and *Festuca rubra* which were present, at low frequency, in the untreated vegetation. It would appear, therefore, that in these various types of vegetation, prior to the addition of fertilizers, the competitive ability of certain component species was held in check by mineral nutrient stress.

Both phenotypic and genetic variation in competitive ability create problems for the ecologist who is attempting to analyse the role of competition in natural vegetation. It is clear, however, that, with care, both types of variation may be recognized and taken into account. What is potentially a more serious difficulty arises from the suggestion, implicit in many ecological writings, that the nature of competition itself may vary fundamentally from one field situation to another so that, relative to other species, a particular species or genotype may be a strong competitor in one site but a weak competitor in another. Inspection of the data which have been quoted in support of this view (e.g. Newman, 1973; Ellenberg and Mueller-Dombois, 1974) shows that, often, this concept has evolved in tandem with loose definitions of competition. This is to say that, in certain cases, shifts in the fortunes of 'competitors' coincident with changes in environment may be attributed more correctly to noncompetitive effects (e.g. differential predation) rather than to any alteration in the relative abilities of the plants to compete for resources.

When noncompetitive effects have been discounted, the remaining objection to a unified concept of competitive ability arises from the fact that competition occurs with respect to several different resources including light, water, various mineral nutrients, and space. Hence it might be supposed that the ability to compete for a given resource varies independently from the ability to compete for each of the others.

There are grounds, both theoretical and empirical, for rejecting this possibility as a major source of variation in competitive ability. It seems more logical to predict that the more usual effect of natural selection will be to cause the abilities to compete for light, water, mineral nutrients, and space to vary in concert and to be developed to a comparable extent in any particular genotype.

In circumstances which allow the rapid development of a large standing crop, the most obvious competition is that occurring above ground with respect to space and light, and it may not be immediately obvious why success should also depend on the effective capture of water and mineral nutrients. However, as pointed out by Mahmoud and Grime (1976), it is clear that rapid production of a large biomass of shoot material, a prerequisite for effective above-ground competition, is dependent upon high rates of uptake of water and mineral nutrients, characteristics which are themselves dependent upon a considerable expenditure of photosynthate on root development. It would appear, therefore, that the abilities to compete for light, mineral nutrients, water, and space are closely inter-dependent. We may suspect, therefore, that, although in productive habitats competition above ground for space and light is more conspicuous, the outcome may be strongly influenced by competition below ground.

The importance of below-ground interactions in competition in productive environments is well illustrated by an experiment conducted by Donald (1958) involving competition between the perennial grasses *Lolium perenne* and *Phalaris tuberosa*. In this pot experiment, a system of partitions, above and below ground, was devised so that it was possible to compare the yield of each of the two species under four treatments, i.e. competition above ground only, competition below ground only, competition above and below ground, and no competition (control). The experiment was carried out at high and low levels of nitrogen supply, but here it is only the former which is relevant to the immediate point at issue. The results of this experiment are presented in Table 3 and show that the clear advantage of *Lolium perenne* over *Phalaris tuberosa* under productive conditions derived from the superior competitive ability of the species both above and below ground.

Competition in productive and unproductive conditions

A criticism which may be levelled at the review of competitive attributes of plants which has been attempted earlier in this chapter is that it refers to characteristics likely to be particularly advantageous during competition in productive, relatively undisturbed vegetation. It may be argued (e.g. Newman, 1973) that elsewhere, and especially in habitats subject to environmental stress, quite different plant characteristics may be advantageous in competition. The crucial test here is to determine whether plants which occur naturally in habitats which are either desiccated, or heavily shaded, or nutrient-depleted are better adapted than plants from productive habitats to compete for specific resources when these are available at low levels of supply.

In the search for evidence bearing on this point, it is convenient first to return to Donald's experiment (Donald, 1958) and, in particular, to consider the results obtained in the low-nitrogen treatments. These results (Table 3) show conclusively that the superior competitive ability exhibited by *Lolium perenne* under productive conditions was maintained when the two grasses competed on nitrogen-deficient soil. However, since both of the grasses in this experiment are characteristic of conditions of moderate to high fertility it may be unwise to draw any general implication from this result.

In an experiment conducted by Mahmoud and Grime (1976), three perennial grasses of contrasted ecology were allowed to compete under productive and unproductive (low nitrogen) conditions. One of the species, *Festuca ovina*, is restricted to unproductive sites, another, *Agrostis tenuis*, is associated with habitats of intermediate fertility, whilst the third, *Arrhenatherum elatius*, is a

Table 3. Measurements of the effects of root and shoot competition between two perennial grasses, *Lolium perenne* and *Phalaris tuberosa*, at high and low rates of nitrogen supply (Donald 1958).

The values tabulated refer to the dry weight of the shoot (g) after 105 days.

	No interspecific competition		Interspecific competition					
			Between shoots		Between roots		Between roots and shoots	
	Lolium	*Phalaris*	*Lolium*	*Phalaris*	*Lolium*	*Phalaris*	*Lolium*	*Phalaris*
High nitrogen	4.71	4.67	4.19	3.19	4.31	1.17	4.72	0.32
Low nitrogen	2.45	2.00	2.71	1.63	2.12	0.35	2.77	0.18

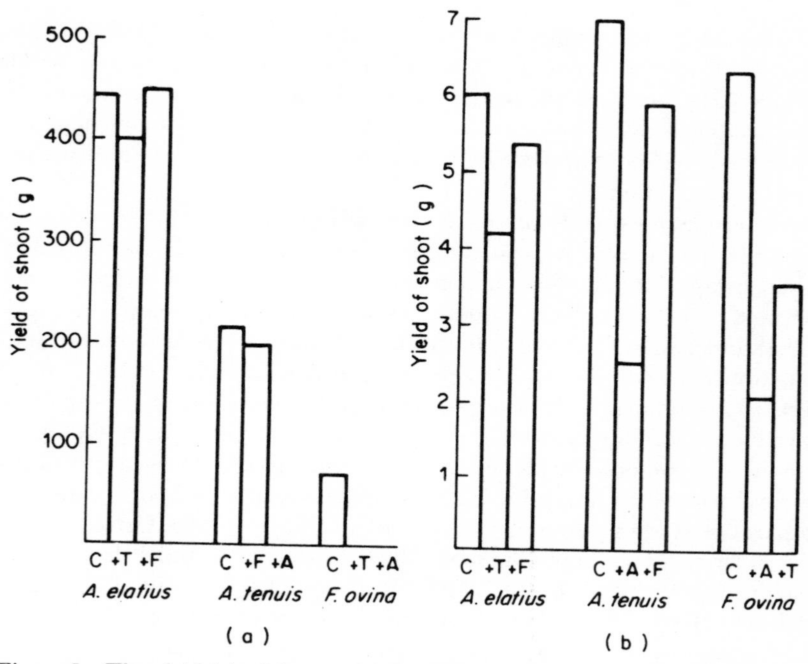

Figure 5 The yield (g) of shoots of *Arrhenatherum elatius* (A), *Agrostis tenuis* (T), and *Festuca ovina* (F) in mixtures (+) and controls (C): (a) under productive conditions (N 176 mg/l), (b) under nitrogen stress (N 5 mg/l) (Mahmoud and Grime, 1976). (Reproduced by permission of Blackwell Scientific Publications Ltd.)

frequent dominant of productive meadows. Each species was grown in monoculture and in separate 1:1 mixtures with each of the other two grasses. The results obtained in the productive treatment are included in Figure 5 and show that the yield of *A. elatius* in monoculture was twice that of *A. tenuis* and approximately seven times that of *F. ovina*. In the mixtures, grown under productive conditions, the interactions between the species were, in each case, conclusive in that *F. ovina* was totally eliminated by both of the other grasses and there were no survivors of *A. tenuis* when this species was grown with *A. elatius*. In the low-nitrogen treatment all three species showed a marked reduction in yield, most pronounced in *A. elatius* and *A. tenuis*. However, there is no evidence in Figure 5 to suggest that the competitive ability of *F. ovina* increased under conditions of nitrogen stress and no impact of *F. ovina* was detected on the yields of either of the other two species.

From this result, it would appear that the ability to compete for nitrogen plays little part in the mechanism whereby *F. ovina* is adapted to survive conditions of low soil fertility. A similar conclusion may be drawn not only with respect to nitrogen but also to other essential mineral nutrients from a number of laboratory experiments (e.g. Bradshaw *et al*., 1964; Hackett, 1965; Clarkson,

1967; Rorison, 1968; Higgs and James, 1969; Fisher *et al.*, 1974; Whelan and Edwards, 1975; McLachlan, 1976; Scaife, 1976) in which plants of infertile habitats have been grown on nutrient-deficient soils and solution-cultures. These studies provide no convincing evidence that plants indigenous to poor soils are more efficient in the uptake of mineral nutrients when these are present at low concentrations.

An alternative explanation for the fitness of plants adapted to conditions of low soil fertility will be presented later (page 28). The essence of this theory is the suggestion that although competition, especially that for water and mineral nutrients, is not restricted to productive habitats, its importance in unproductive habitats is small relative to the ability to conserve the resources which have been captured and to resist the severe hazards to survival which characterize infertile environments. This argument may be extended to the full range of unproductive habitats including those in which light and water supply are the main limiting factors. Further evidence of the decline in importance of competition for resources in unproductive vegetation is available from a wide range of comparative studies (Ashton, 1958; Pigott and Taylor, 1964; Grime and Jeffrey, 1965; Grime, 1965, 1966; Clarkson, 1967; Myerscough and Whitehead, 1967; Hutchinson, 1967; Loach, 1967, 1970; Rorison, 1968; Parsons, 1968a, b; Higgs and James, 1969). As explained later (page 50), it would appear that competitive characteristics such as rapid potential growth-rate and high phenotypic plasticity with respect to the development of photosynthate (e.g. increase in leaf area in shade and increase in root : shoot ratio under mineral nutritional stress) become disadvantageous in circumstances of extreme and more or less continuous environmental stress.

General features of competitors

Phenology and phenotypic plasticity

From the evidence and arguments which have been presented in this chapter it seems reasonable to conclude that high competitive ability is recognizable as a family of genetic characteristics which permit a high rate of acquisition of resources in productive, crowded vegetation. Under these conditions, natural selection appears to favour those plants which are best equipped both to tap the surplus of resources above and below ground and to maximize dry matter production. In this respect two competitive characteristics are of particular importance. The first consists of the potential to produce a dense canopy of leaves and a large root surface area during the period in the year when conditions are most favourable to high productivity. The second is the capacity for rapid morphogenetic adjustments both in the apportionment of photosynthate between root and shoot and in the size, morphology, and distribution of individual leaves (Blackman and Wilson, 1951a, b; Blackman and Black, 1959; Grime and Jeffrey, 1965) and roots (Drew, Saker, and Ashley, 1973), a characteristic which involves a high rate of reinvestment of captured resources in growth and in respiration. The effect of such responses, coupled with the rapid turnover of the leaves and roots, is to bring about, during the growing season, a constant readjustment of the spatial distribution of the absorptive surfaces (i.e. the leaf canopy and root surface area) of the plant. The advantage which the competitor appears to derive from this high flexibility is the potential to respond rapidly to changes in the distribution of resources within the habitat. On page 46 this 'foraging' mechanism of the competitor is contrasted with the very different mechanism of resource-capture which characterizes the stress-tolerator.

Life-form and ecology

Until growth analysis experiments and competition studies have been conducted on a wide variety of species from contrasted habitats it may not be possible to recognize with certainty the life-form classes and vegetation types which include species of high competitive ability. However, some tentative conclusions may be drawn with regard to the relationship between competitive ability and life-form. In particular it is evident, from the limited amount of data which is available, that no broad distinction with respect to competitive ability may be drawn between herbaceous and woody species. Perennial herbs, shrubs, and trees each appear to encompass a wide range of competitive abilities. As explained in Chapter 5, this observation is of considerable significance in relation to the process of vegetation succession.

STRESS-TOLERATORS

A definition of stress

Dry matter production in vegetation is subject to a variety of environmental constraints, the most frequent of which are related to shortages and excesses in the supply of solar energy, water, and mineral nutrients. Plant species and even different genotypes may differ in susceptibility to particular forms of stress and, in consequence, each may exercise a different effect upon vegetation composition. Because, over the course of a year, several stresses may operate intermittently in the same habitat, analyses of the impact of stress may become quite complex.

A further complication arises from the fact that certain stresses either originate from or are intensified by the vegetation itself. Among the most important types of plant-induced stress are those arising from shading and depletion of the levels of mineral nutrients in the soil following their accumulation in the plant biomass. In addition, growth-inhibitors may be released into the soil either by direct secretion or as a result of microbial decay of plant residues.

In order to accommodate its diverse forms, stress will be defined simply as *the external constraints which limit the rate of dry matter production of all or part of the vegetation*.

Stress in productive and unproductive habitats

Identification of the forms of stress characteristic of particular habitats is but one aspect of the analysis which is necessary in order to determine the influence of stress upon vegetation. Another requirement is to estimate the extent to which stress is limiting primary production in different types of vegetation. The need for such studies arises from the fact that the role of stress changes according to the productivity of the environment.

In productive, undisturbed habitats the vegetation is composed of potentially-large, fast-growing plants, and stress arises mainly as a consequence of local or temporary depletion of resources by competitors.* In these circumstances, as we have seen earlier (pages 13 and 20), natural selection is likely to favour plants in which exposure to stress, whether plant-induced or imposed directly by the environment, induces rapid morphogenetic responses which continue to maximize the capture of resources and the production of dry matter. However, we may suspect that such responses to stress may have little survival value under conditions in which severe stress in one form or another is a constant feature of the environment.

In chronically unproductive habitats therefore, there is little opportunity for the phenology or morphogenetic responses of the plant to provide mechanisms of stress-avoidance and the most conspicuous effect of severe and continuous stress is to eliminate or to debilitate species of high competitive ability and to cause them to be replaced by plants which are capable of tolerating the prevailing forms of stress. The stress-tolerators comprise an extremely diverse assortment of plants which on first inspection appear to be far too varied in life-form and in ecology to be included in the same category. However, the morphological and taxonomic diversity among the stress-tolerators belies the conformity of life-history and physiology whereby these plants are adapted to survive in continuously unproductive conditions. This is to suggest that although the various types of stress-tolerators differ through the possession of mechanisms adapted to the specific forms of stress operating in their habitats, all exhibit a suite of characteristics necessary for survival in conditions in which the level of production remains consistently low. Some of the evidence which supports this hypothesis will now be examined by considering briefly adaptation to severe stress in four contrasted types of habitat. In two of these (arctic–alpine and arid habitats) the plant biomass is small and stress is mainly imposed by the environment. In the third (shaded habitats) stress is plant-induced, whilst in the fourth (nutrient-deficient habitats) stress may be due to the low fertility of the habitat or to accumulation of the mineral nutrients in the vegetation or to a combination of the two.

*More precisely, there is a positive feedback between the ability of the stronger competitors to capture resources and their tendency to subject weaker competitors to 'plant-induced' or 'plant-intensified' stresses, principally by shading and by depletion of water or mineral nutrients in the rhizosphere. This phenomenon may be described as competitive dominance and is examined further on page 125.

Tolerance of severe stress in various types of habitat

Stress-tolerance in arctic and alpine habitats

Without doubt, the dominant environmental stress in arctic and alpine habitats is low temperature. In these habitats, the opportunity for growth is limited to a short summer season. For the remainder of the year growth is prevented by low temperature and the vegetation is either covered by snow or, where it remains exposed on ridges, is subjected to extreme cold coupled with the desiccating effect of dry winds. During the growing season itself, production is often severely restricted not only by low temperatures but also by desiccation and mineral nutrient stress (Haag, 1974), the latter arising largely as a result of the low microbial activity of the soil. Alpine vegetation types are subject to additional stresses peculiar to high altitudes, including strong winds and intense solar radiation.

The adaptive characteristics of the terrestrial plants of arctic and alpine habitats have been examined in excellent reviews by Billings and Mooney (1968) and Bliss (1971), and here comment will be restricted to certain of the most generally occurring adaptations.

The predominant life-forms in arctic and alpine vegetation are very low-growing evergreen shrubs, small perennial herbs, bryophytes, and lichens. The adaptive significance of the small stature of all of these plants appears to be related, in part, to the observation that during the winter when the ground is frozen and no water is available to the roots the aerial parts of low-growing plants tend to be insulated from desiccation by a covering of snow. However, as suggested by Boysen-Jensen (1932), quite apart from the risk of winter-kill of shoots projecting through the snow cover, the absence of larger plants and, in particular, erect shrubs and trees, is also related to the fact that the productivity of tundra and alpine habitats is so low that it is unlikely that there will be a surplus of photosynthate sufficient to sustain either wood production or the annual turnover of dry matter characteristic of deciduous trees and tall herbs.

A conspicuous feature of arctic and alpine floras is the preponderance of evergreen species (Polunin, 1948). Annual plants are extremely rare and the majority of bryophytes, lichens, herbs, and small shrubs remain green throughout the year. The advantage of the evergreen habit is particularly clear in long-lived prostrate shrubs. Here, Billings and Mooney (1968) suggest that the main advantage of the evergreen habit is that it obviates the necessity 'to spend food resources on a wholly new photosynthetic apparatus each year.' In this connection, it is interesting to note that Hadley and Bliss (1964) found that evergreen tundra shrubs had lower photosynthetic rates than the deciduous shrub *Vaccinium uliginosum* and various herbaceous species. The ability to survive, despite these lower rates, is apparently due to the longer functional life of individual leaves. Hadley and Bliss (1964) also suggest that the older leaves may act as winter food storage organs since lipids and proteins are mobilized and translocated from old to new leaves during the growing season. Billings and Mooney (1968) note that evergreen shrubs in alpine conditions tend to break bud dormancy later than deciduous species exposed to the same conditions and suggest that evergreens 'can afford the apparent waste of these days of uncertain weather early in the growing season since their older leaves are already in photosynthetic operation.'

The importance of the evergreen habit in arctic vegetation has been exemplified more recently by a detailed analysis (Callaghan 1976) of the life-cycle of the sedge *Carex bigelowii* in which it has been found that the life-span of individual tillers in arctic populations frequently extends to four years.

It would appear, from the relatively small amount of information which is available, that seed production in arctic and alpine habitats is erratic and hazardous. However, as Billings and Mooney (1968) point out, this is compensated by the fact that the majority of species are long-lived perennials, many of which are capable of vegetative reproduction (see page 81).

Stress-tolerance in arid habitats

The comprehensive reviews by authors such as Walter (1973), Slatyer (1967), and Levitt (1975) make clear the pitfalls in any attempt to generalize about the ways in which plants are adapted to exploit conditions of low annual rainfall. However, if attention is confined to those species which, in their natural habitats, experience long periods of desiccating conditions without access to underground reservoirs of soil moisture, common adaptive features are apparent. These xerophytes are perennials in which the vegetative plant is adapted to survive for extended periods during which little water is available. Several types of xerophytes may be distinguished according to the severity of moisture stress which they can survive.

In habitats which experience a short annual wet season the most commonly occurring xerophytes are the sclerophylls (Grieve, 1956; Monk, 1966). This group includes both small shrubs and trees, such as the evergreen oaks and the olive, all of which are distinguished by the possession of small, hard leaves which are retained throughout the dry season. It has been established (Gates, 1968) that under conditions of high radiation load and restriction of transpiration by stomatal closure, small leaves dissipate heat more efficiently than large ones. Perhaps even more important is the suggestion of Walter (1973) that the ecological advantage of sclerophylly is related to the ability of sclerophyllous species 'to conduct active gaseous exchange in the presence of an adequate water supply but to cut it down radically by shutting the stomata when water is scarce,' a mechanism which enables these plants to 'survive months of drought with neither alteration in plasma hydrature nor reduction in leaf area' and, when rains occur, to 'immediately take up production again.'

Under severely desiccating conditions, the most persistent of the xerophytes are the succulents. These are plants in which water is stored in swollen leaves, stems, or roots. During drought periods no water absorption occurs, but following rain, small short-lived roots may be produced extremely rapidly. Succulents are also distinguished by peculiarities in stomatal physiology and metabolism.

The stomata open at night and remain tightly closed during the day. Gaseous exchange occurs during the period of stomatal opening, at which time carbon dioxide is incorporated into organic acids. These are decarboxylated in the daylight to release carbon dioxide for photosynthesis. This mechanism, known as crassulacean acid-metabolism (Thomas, 1949) appears to represent a mechanism whereby low levels of photosynthesis may be maintained with minimal transpiration.

A well-known feature of many xerophytes and, in particular, the succulents, is the rarity or erratic nature of flowering.

Stress-tolerance in shaded habitats

On superficial acquaintance, the stresses associated with shade appear to differ from those considered under the two previous headings in that they are not directly attributable to gross features of climate. However, although shade itself is not imposed by the physical environment it becomes important only in climatic regimes which are conducive to the development of dense canopies. It may be important to bear in mind, therefore, that shade at its greatest intensities frequently coincides (1) with the high temperatures and humidities of tropical and subtropical climates or with the warm summer conditions of temperate regions and (2) with conditions of mineral nutrient depletion associated with the development of a large plant biomass. Hence, although the discussion which follows refers to adaptations to shade, there is a strong probability that some of the plant characteristics described are related to co-tolerance of shade, warm temperatures, and mineral nutrient stress.

The intensity of shade experienced near the ground surface depends upon the number of layers of foliage present and upon the light absorbing and reflecting characteristics of the canopy. Although the amount of light intercepted by a dense community of herbaceous species may be comparable with that intercepted by forest (Monsi and Saeki, 1953), there is, of course, a major difference with respect to the height of the shaded stratum. Within herbaceous vegetation, the shaded stratum is low and all or part of it is renewed annually by extension of shoots and individual leaves from positions near the ground. In forest, however, the shaded stratum is high and arises by expansion of foliage *in situ*.

In dense herbaceous vegetation, whether in the open or in forest clearings, small differences in height above ground are associated with large changes in intensity, direction, and quality of radiation and success among the component herbs and tree seedlings may depend to a considerable extent upon the ability to compete for light and to project leaves into the higher light intensites above the herb layer. In contrast, beneath dense tree canopies (and, more especially, those composed of evergreen species), herbaceous vegetation is usually sparse and, in consequence, vertical gradients in light intensity are less pronounced near the forest floor. Here, the ability to compete for light is likely to be secondary in importance to the capacity to tolerate shade conditions. The need to distinguish between competition for light and tolerance of shade becomes clear when an attempt is made to analyse the results of shade experiments.

The effect of shading upon growth-rate and morphogenesis has been examined in a large number of experiments by growing plants, usually as seedlings, under screens of cotton, plastic, or metal (e.g. Burns, 1923; Blackman and Rutter, 1948; Blackman and Wilson, 1951a, b; Bourdeau and Laverick, 1958; Blackman and Black, 1959; Grime and Jeffrey, 1965; Björkman and Holmgren, 1963, 1966; Loach, 1970). The general conclusions which may be drawn from these experiments are (1) that, in response to shade, the majority of plants produce less dry matter, retain photosynthate in the shoot at the expense of root growth, develop longer internodes and petioles, and produce larger, thinner leaves, and (2) that species differ very considerably both with respect to the magnitude and to the rate of these various responses. However, when the responses of different ecological groups of plants are compared, a paradox becomes apparent. This arises from the fact that the capacity to maximize dry matter production in shade through modification of the phenotype is most apparent in species characteristic of unshaded or lightly-shaded environments whilst plants associated with deep shade tend to grow slowly and to show much less pronounced morphogenetic responses to shade treatment. An examination of this difference is available from the investigations of Loach (1967, 1970) who used screens to compare the shade responses of several North American trees. These results (Figure 6) establish quite clearly that morphological plasticity in shade is most apparent in species such as *Liriodendron tulipifera* which normally colonizes unshaded habitats or woodland clearings, whilst shade-tolerant species, such as *Fagus grandifolia*, exhibit a much smaller degree of phenotypic modification.

From these results it would appear that the rapidity of phenotypic response and the comparatively high growth rates in shade of species such as *L. tulipifera* are attributes which allow seedlings of these species to compete for light in rapidly-expanding herbaceous vegetation such as that which occurs in forest clearings. The low rates of growth and the small extent of phenotypic response to shading in the shade-tolerant trees suggest that adaptation to shade in these plants may be concerned more with the ability to survive for extended periods in deep shade than with the capacity to maximize light interception and dry matter production. A similar conclusion is prompted by the observation that many of the most shade-tolerant herbaceous plants, e.g. *Pachysandra* spp., *Deschampsia flexuosa*, have morphologies which allow considerable self-shading.

In a number of comparative experiments (Grime, 1965; Loach, 1970; Grime and Hunt, 1975) shade-tolerant herbs and tree seedlings have been found to exhibit consistently slow relative growth-rates under conditions in which plants characteristic of productive habitats grew extemely rapidly. The results of shade-screen experiments with tree seedlings (Loach, 1970) and with the shade-tolerant grass *Deschampsia flexuosa* (Mahmoud and Grime, 1974) indicate that these low relative growth-rates are genetically determined in that they are maintained at both high and low light intensities.

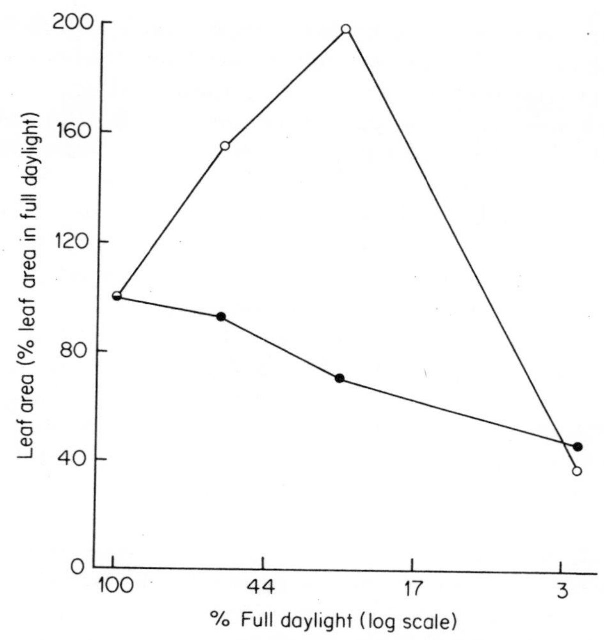

Figure 6 Leaf area as a function of shade treatment over a period of thirteen weeks in seedlings of a shade-tolerant tree, *Fagus grandifolia* (•–•), and a shade-intolerant species, *Liriodendron tulipifera* (o–o) (Loach, 1970). (Reproduced by permission of Blackwell Scientific Publications Ltd.)

Measurements in the dark on leaves from a variety of species (Bordeau and Laverick, 1958; Björkman and Holmgren, 1963, 1966; Grime, 1965; Loach, 1970, Taylor and Pearcy, 1976) suggest that shade-tolerant species have comparatively low respiratory rates and that these tend to remain low in shade-grown plants. It is not clear whether these low rates of respiration are imposed biochemically or arise through reduced permeability of the leaf. However, experiments by Woods and Turner (1971) have shown that stomatal responses to changes in light intensity in a range of North American trees are consistently more rapid in shade-tolerant species. This most interesting observation suggests two avenues for speculation with regard to the physiology of shade tolerance. The first concerns the possibility that following stomatal closure in darkness and at low light intensities depletion of respiratory substrates may be inhibited in some way by reduced ventilation of the leaf. The second is that rapid opening of stomata may allow shaded leaves to exploit brief periods of illumination due to sun flecks.

These physiological studies suggest that natural selection in deeply-shaded habitats has been associated with the evolution of mechanisms of conserving energy rather than with those which increase the quantity of energy captured.

In particular, it seems likely that low respiratory rates may be important in maintaining the carbon balance of plants exposed simultaneously to low light intensity and high temperature. Experiments in which plants have been grown in various combinations of light and temperature (Grime, 1966; Mason, 1976) suggest that in some shade-tolerant species respiration is relatively insensitive to changes in temperature.

An additional mechanism whereby carbon and energy may be conserved in shade-tolerant plants is evident from the work of Björkman (1968a, b) who has demonstrated that the lower photosynthetic rates of shade-adapted ecotypes of the herbaceous perennial *Solidago virgaurea* is related to the lower carboxylating enzyme content of the leaves. This evidence has prompted Mooney (1972) to observe that 'since a major portion of the leaf protein is carboxydismutase, there can be a considerable conservation of carbon by making less of this enzyme in light-limited habitats, where it would be of little advantage.' Consistent with this hypothesis are the results of a more recent investigation by Taylor and Pearcy (1976) in which it was found that in a number of North American shade-tolerant herbs including *Trillium grandiflorum*, *Podophyllum peltatum*, and *Solidago flexicaulis*, shoots developed under shaded conditions showed marked reductions in carboxylating capacity whereas species such as *Erythronium americanum* and *Allium tricoccum*, which are vernal shade-intolerant species, showed no evidence of such flexibility.

Among the trees, climbing plants, and epiphytes which occupy the shaded stratum of tropical forests, evergreen species predominate. In temperate woodlands also, the most shade-tolerant herbaceous plants tend to remain green throughout the year (Kubiček and Brechtl, 1970; Hughes, 1975). In the British Isles, for example, deciduous woodlands contain a variety of slow-growing evergreen species such as *Galeobdolon luteum*, *Milium effusum*, *Deschampsia flexuosa*, and *Veronica montana*.

A characteristic of many shade-tolerant species is the paucity of flowering and seed production under heavily shaded conditions. This phenomenon is particularly obvious in British woodlands where flowering in many common shade plants such as *Hedera helix*, *Lonicera periclymenum*, and *Rubus fruticosus* agg. is usually restricted to plants exposed to sunlight at the margins of woods or beneath gaps in the tree canopy.

Stress-tolerance in nutrient deficient habitats

Under this heading attention will be confined to habitats in which mineral nutrient stress arises from the impoverished nature of the habitat. This is because it is in these habitats that mechanisms of adaptation to mineral nutrient stress have been most closely studied. However, it is vital to the arguments developed in Chapter 5 to recognize that severe stress may also arise under conditions in which mineral nutrients are sequestered in the living or non-living parts of the biomass.

Since the incursion of experimental methods into agricultural and ecological research, considerable effort has been devoted to the task of identifying the stress factors which cause particular soils to be infertile. The types of naturally-occurring infertile soils which have been intensively studied include those which are acidic, calcareous, or derived from serpentine rock. Industrial spoils have been investigated also and particular attention has been paid to coal mine waste and spoil contaminated by heavy metals such as lead, copper, and zinc.

In the majority of habitats which have been examined, low soil fertility has been found to be associated with several different forms of stress operating simultaneously or in seasonal succession. The complexity possible in an analysis of the causes of soil infertility may be illustrated by reference to the list (Table 4) composed by Hewitt (1952) of the stresses which may contribute to low productivity on highly acidic soils. A similar degree of complexity is suspected in the mechanism inhibiting vegetation development on serpentine soils. Walker (1954), for example, states that, in order to survive in serpentine habitats, plants must be 'tolerant of low calcium levels and, in addition, be tolerant of one or more of the following: high concentrations of chromium and nickel, high magnesium, low levels of major nutrients, low available molybdenum, drought, and other undesirable aspects of shallow stony ground.'

Before any progress can be made in an attempt to generalize about the way in which vegetation is adapted to conditions of low soil fertility it is necessary to

Table 4. List of the main factors contributing to low productivity on highly acidic soils (Hewitt 1952).

(1) Direct injury by hydrogen ions (low pH).

(2) Indirect effects of low pH.
- (a) Physiologically impaired absorption of calcium, magnesium and phosphorus.
- (b) Increased solubility, to a toxic extent, of aluminium, manganese, and possibly iron and heavy metals.
- (c) Reduced availability of phosphorus partly by interaction with aluminium or iron, possibly after absorption.
- (d) Reduced availability of molybdenum.

(3) Low base status.
- (a) Calcium deficiency.
- (b) Deficiencies of magnesium, potassium or possibly sodium.

(4) Abnormal biotic factors.
- (a) Impaired nitrogen cycle and nitrogen fixation.
- (b) Impaired mycorrhizal activity.
- (c) Increased attack by certain soil pathogens.

(5) Accumulation of soil organic acids or other toxic compounds due to unfavourable oxidation-reduction conditions.

distinguish between those soil characteristics which are common to most infertile habitats and those which are peculiar to specific types of soils. When a survey is made of the extensive literature concerning various forms of soil infertility, the most consistent components of the 'infertility complex' are major nutrient deficiencies and, in particular, those of phosphorus and nitrogen.* Accordingly, the remainder of this section is concerned with an examination of the adaptations of vegetation to conditions of major nutrient deficiency.

When the range of vegetation types characteristic of severely nutrient deficient soils are surveyed, certain common features are immediately apparent. Although the identity of the species involved varies in different parts of the world and according to local factors such as soil type and vegetation management, similar plant morphologies may be recognized. The herbaceous component shows a marked reduction in growth form. Among the grasses, narrow-leaved tussock forms predominate and a high proportion of the dicotyledons are creeping or rosette species. Under management conditions which allow the development of woody vegetation, nutrient-deficient soils are usually colonized by small (often coniferous) trees or sclerophyllous shrubs (Mason, 1946; Gardner and Bradshaw, 1954; McMillan, 1956; Monk, 1966; van Steenis, 1972; Vogl, 1973). In Europe and North America, members of the Ericaceae such as *Calluna vulgaris*, *Vaccinium myrtillus*, and *Erica cinerea* are particularly common on highly acidic soils, whilst on shallow nutrient-deficient calcareous soils species such as *Helianthemum chamaecistus* and *Thymus drucei* occur. The xeromorphy evident in all these shrubs could be interpreted to be an adaptation to winter or summer desiccation. However, this explanation cannot be applied in the case of the xerophytic shrubs which are known to occur on infertile soils in tropical and subtropical regions. In New Caledonia, for example, Birrel and Wright (1945) described xerophyllous shrub vegetation varying between 1–2 m in height on serpentine soil and commented upon its 'unusual appearance in this region of high rainfall in which tropical forest is the ordinary plant cover.'

Another feature which is characteristic of the vegetation of nutrient-deficient soils is the high frequency of species of inherently slow growth-rate. Among the first to recognize this was Kruckeberg (1954) who noted that, even when grown on fertile soils, certain ecotypes of herbaceous plants adapted to survive on serpentine soils in North America grew slowly in comparison with species and ecotypes from fertile habitats. Another important early contribution was that of Beadle (1954, 1962) who recognized that slow growth-rates were characteristic of species of *Eucalyptus* growing on Australian soils with low phosphorus availability. Slow rates of growth were also found to occur in *Festuca ovina* and *Nardus stricta*, two grasses of widespread occurrence on infertile soils in the British Isles (Bradshaw *et al.*, 1964) and Jowett (1964) concluded from his

* Deficiencies of soil phosphorus arise through a variety of mechanisms including scarcity in the parent rocks, precipitation of insoluble salts of calcium, aluminium, and iron, and sequestration by higher plants and soil micro-organisms. The most common causes of nitrogen deficiency are various forms of inhibition of nitrifying and nitrogen-fixing micro-organisms.

investigation with the grass *Agrostis tenuis* that the effect of natural selection under conditions of severe mineral nutrient-deficiency had been to reduce the potential growth rate of local populations established on mine-waste in Wales. Subsequent investigations (e.g. Hackett, 1965; Clarkson, 1967; Higgs and James, 1969; Grime and Hunt, 1975) have confirmed that there is a strong correlation between low potential growth-rate and tolerance of mineral nutrient deficiencies.

It would appear, therefore, that adaptation for survival on infertile soils has involved, in both woody and herbaceous species, reductions in stature, in leaf form, and in potential growth-rate. The explanation which has been put forward, with varying degrees of elaboration, by a number of authors (e.g. Kruckeberg, 1954; Loveless, 1961; Jowett, 1964) to account for this phenomenon is that it is primarily an adaptation for survival under conditions of low mineral nutrient supply. This is to suggest that, under conditions in which elements such as phosphorus and nitrogen are scarcely available, natural selection has led to the evolution of plant species and ecotypes which make low demands upon the mineral nutrient reserves of the soil. Consistent with this hypothesis are the results of a number of experiments in which species from infertile habitats have been grown under various levels of supply of major mineral nutrients (Bradshaw *et al.*, 1964; Hackett, 1965; Clarkson, 1967). From these studies there is no convincing evidence that species normally restricted to infertile soils are better adapted than species from fertile habitats to maintain dry matter production under conditions in which mineral nutrients are provided at low rates of supply. In Figure 7, for example, the results of an experiment by Bradshaw *et al.* (1964) illustrate clearly that, under low levels of nitrate nitrogen supply, *Festuca ovina* and *Nardus stricta*, both grasses of infertile pastures, are outyielded to a considerable extent by *Lolium perenne* and *Cynosurus cristatus*, species which are normally restricted to fertile soils.

An advantage which a low rate of dry matter production confers upon a plant growing on an infertile soil is that, during periods of the year when mineral nutrients are more readily available, uptake is likely to exceed the rate of utilization in growth, allowing reserves to accumulate. An example of this 'luxury uptake' which presumably may benefit the plant during subsequent periods of nutrient shortage is provided in Figure 8 which is taken from an experiment by Clarkson (1967) and shows the concentration of phosphorus accumulated in the plant tissues by two species of *Agrostis* grown at various levels of phosphorus supply. The data show that in *A. setacea*, a slow-growing species of infertile acidic grasslands, there was a progressive accumulation of phosphorus within the plant as the concentration of the element increased in the external medium. In contrast, *A. stolonifera*, because it responded to increasing levels of phosphorus by growing more rapidly, did not accumulate phosphorus to a level surplus to the growth requirements of the plant.

Whilst the capacity of slow-growing plants of infertile soils to accumulate high internal concentrations of nutrient elements may be of considerable survival value in the natural habitats of these species, there is evidence which

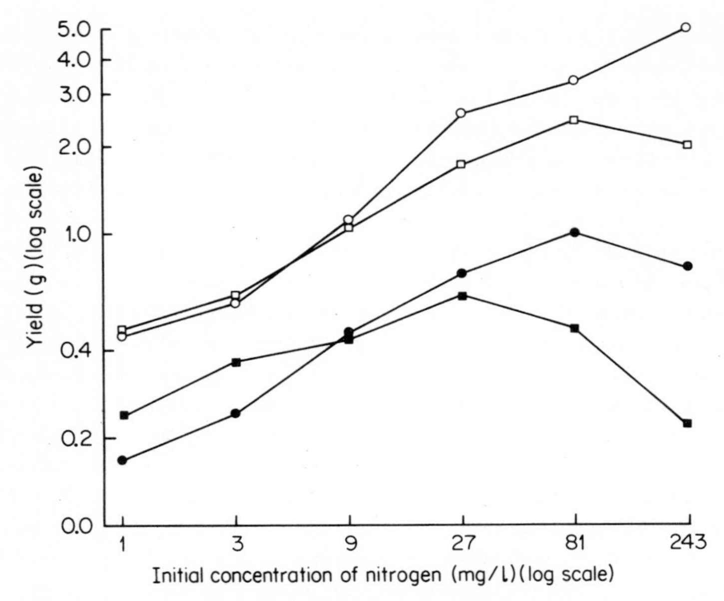

Figure 7 The yield of four grasses grown for eight weeks in unreplenished sand cultures provided with various initial concentrations of nitrogen. Two of the species, *Lolium perenne* (○–○) and *Cynosurus cristatus* (□–□), are characteristic of fertile pastures whilst the remaining pair, *Festuca ovina* (●–●) and *Nardus stricta* (■–■), are species of infertile grassland (Bradshaw *et al*., 1964). (Reproduced by permission of Blackwell Scientific Publications Ltd.)

suggests that this characteristic may be disadvantageous when these same plants colonize more fertile soils. In a number of experiments (Bradshaw *et al*., 1964; Jeffries and Willis, 1964; Ingestad, 1973; Jones, 1974) in which plants associated with nutrient-deficient habitats were supplied with high rates of mineral nutrients, elements appear to have been accumulated in quantities which were detrimental to the growth of the plants.

An additional feature of many of the slow-growing species characteristic of infertile soils is the lack of a sharply defined seasonal variation in shoot biomass (Figure 9). The majority of both the woody and the herbaceous species characteristic of infertile habitats are evergreen plants in which the leaves have a comparatively long life-span. Measurement by Williamson (1976), for example, on grasses of calcareous pastures in Southern England have shown that in the evergreen species of nutrient-deficient soils, *Helictotrichon pratense* and *Helictotrichon pubescens*, the functional life of the leaf (Figure 10) is considerably longer than in *Arrhenatherum elatius* and *Dactylis glomerata* species which are usually associated with more fertile soil conditions and show a well-defined summer peak in shoot biomass. A consequence of the greater longevity and slower replacement of leaves in plants of infertile habitats is a reduction in the rate of nutrient cycling between plant and soil and a smaller risk of loss of

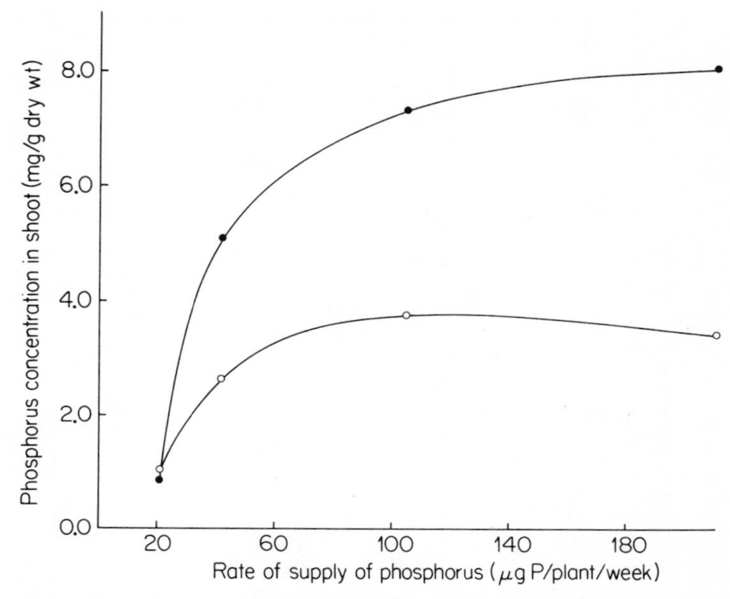

Figure 8 Measurements of the concentration of phosphorus in two perennial grasses grown for a period of six weeks at various rates of phosphorus supply. One species, *Agrostis stolonifera* (○–○), is associated with soils of high fertility whilst the other, *Agrostis setacea* (●–●), is restricted to infertile acidic soils (Clarkson, 1967). (Reproduced by permission of Blackwell Scientific Publications Ltd.)

mineral nutrients either by leaching or by incorporation into other organisms exploiting the habitat (Monk, 1966, 1971; Thomas and Grigal, 1976). This observation is consistent with the general conclusion that plants of infertile habitats are adapted to conserve mineral nutrients rather than to maximize the rate of capture.

General features of stress-tolerance

From the preceding survey it is apparent that vascular plants adapted to contrasted forms of severe stress are comparatively long-lived and exhibit a range of features which, although varying in detailed mechanism, represent basically similar adaptations for endurance of conditions of limited productivity. These features, which are listed in Table 6, include inherently slow rates of growth, the evergreen habit, long-lived organs, sequestration and slow turnover of carbon, mineral nutrients, and water, infrequent flowering, and the presence of mechanisms which allow the intake of resources during temporarily favourable conditions. The latter consist not only of the presence throughout the year of functional leaves and, probably, roots also (Jeffrey, 1967; Chapin and Bloom, 1976; Thomas and Grigal, 1976) but, in addition, special mechanisms such as

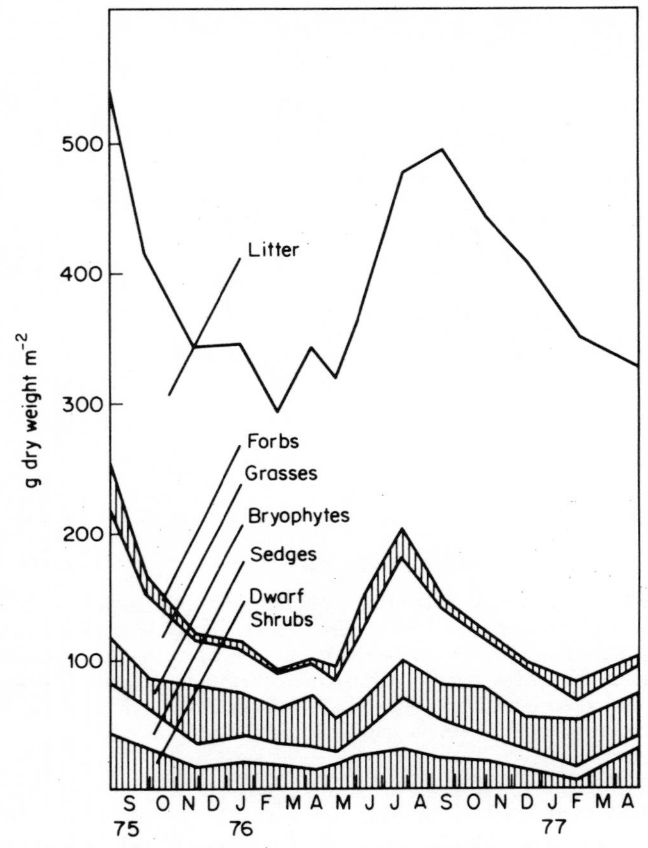

Figure 9 Seasonal variation in the composition of the standing crop in an unfertilized limestone pasture in Northern England. A feature of the figure is the lack of pronounced seasonal changes in the contribution made by any of the five life-form classes, a consequence of the high frequency of slow-growing evergreen species (Furness, 1978).

the rapid activating of stomata in scelerophylls and shade plants and the rapid sprouting of roots by succulents.

A feature consistently associated with the stress-tolerator is low morphogenetic plasticity. In terms of the growth physiology of these plants this characteristic is not difficult to understand. Growth in stress-tolerators occurs intermittently, and for most of the time, therefore, differentiating (i.e. potentially responsive) tissue forms a very small proportion of the biomass.

In stress-tolerators the most important responses to environmental variation are physiological rather than morphogenetic. Although research on this subject is at an early stage of development it is now apparent that the long functional life of individual shoots and roots in many stress-tolerators is charac-

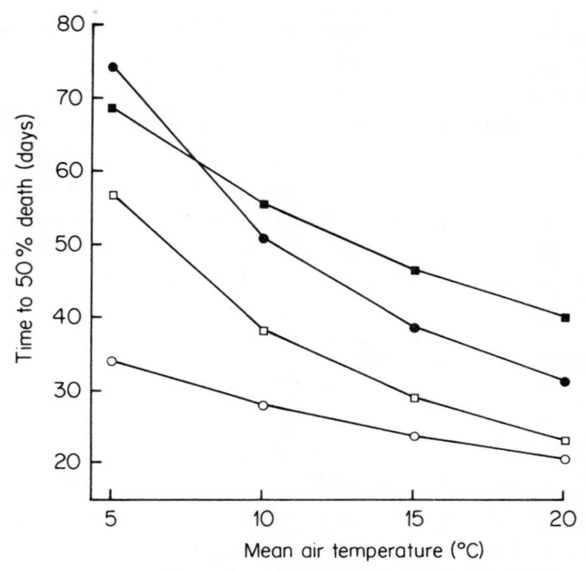

Figure 10 Estimations of response to temperature in the death rate of the leaves of four grasses established in the same derelict grassland in Southern England. Two of the species examined, *Arrhenatherum elatius* (○–○) and *Dactylis glomerata* (□–□), are usually associated with conditions of moderately high soil fertility. The two remaining species, *Helictotrichon pratense* (●–●) and *H. pubescens* (■–■), are more commonly found on infertile soils (Williamson, 1976). (Reproduced by permission of Blackwell Scientific Publications Ltd.)

terized by seasonal or short-term changes which maintain their viability and functional efficiency under changing environmental conditions. Reference has been made already (page 28) to evidence of the capacity of some woodland plants to modify the size of the carboxylating system in response to shade conditions. Changes in the photosynthetic apparatus are also apparent in the temperature acclimation and seasonal variation in efficiency of carbon assimilation described for vascular plants in desert and temperate localities (Mooney and West, 1964; Strain and Chase, 1966; Mooney and Shropshire, 1967; Mooney and Harrison, 1970) and in arctic, alpine, and subarctic species (Billings *et al*., 1971; Grace and Woolhouse, 1970; Billings and Godfrey, 1968). Physiological changes corresponding to fluctuations in seasonal conditions have been reported also for temperate forest species (Büttner, 1971), bryophytes (Hosakawa *et al*., 1964; Miyata and Hosakawa, 1961; Oechel and Collins, 1973; Hicklenton and Oechel, 1976), and arctic lichens (Larsen and Kershaw, 1975). It is interesting to note that acclimation to temperature may be extremely rapid in certain species. Mooney and Shropshire (1967) have shown that in coastal populations of the Californian desert shrub *Encelia californica* changes in photosynthetic response to temperature may be induced within a period of twenty-four hours, and a similar phenomenon has been described (Hicklenton and Oechel, 1976) in the subarctic moss *Dicranum fuscescens*.

As in the case of the competitive strategy, no simple generalization can be made with regard to the stature and life-form of stress-tolerant plants. Although, in extremely unproductive environments stress-tolerance is associated with trees, shrubs, and herbs of reduced stature, account must be taken of the fact that many of the shade-tolerant trees in temperate and tropical habitats are long-lived and attain a comparatively large size at maturity.

Stress-tolerance and symbiosis

Lichens

Four features which, under the last heading, have been associated with stress-tolerance in flowering plants (slow growth-rate, longevity, opportunism, and physiological acclimation) are expressed in a most extreme form in lichens. This is perhaps hardly surprising in view of their ecology. Lichens are able to survive in extremely harsh environments under conditions in which vascular plants may be totally excluded and in which they experience extremes of temperature and moisture supply and are subject to low availability in mineral nutrients. Although there is some diversity of opinion with regard to the longevity of lichens (see, for example, the review by Billings and Mooney, 1968) there is general agreement that many are exceedingly long-lived. From experimental studies, such as those of Farrar (1976a, b, c), there is abundant evidence of the tendency of lichens to sequester a high proportion of the photosynthate rather than to expend it in growth. Much of the assimilate appears to be stored as sugar-alcohols (polyols) in the fungal component. Lichens are able both to remain alive during prolonged periods of desiccation and, on rewetting, to resume nutrient uptake and photosynthesis extremely rapidly. The studies of Kershaw and his colleagues on Canadian populations of *Peltigera canina* and *P. polydactyla* (Kershaw 1977a, b; MacFarlane and Kershaw, 1977) provide an ample illustration of the capacity of lichens for rapid seasonal acclimation.

Ectotrophic mycorrhizas

An interesting parallel may be explored between the lichens and another type of symbiosis involving fungi. In many trees and shrubs the roots develop a thick investment of fungal hyphae (ectotrophic mycorrhizas) which are connected on the inside with living cells of the root and on the outside with hyphae which are in contact with the soil and leaf litter.

A considerable amount of physiological work has been carried out in order to assess the ecological significance of ectotrophic mycorrhizal associations, and the results have been reviewed extensively (Harley, 1969, 1970, 1971). It has been clearly established that the uptake of mineral nutrients such as phosphorus and nitrogen may be facilitated, and the yield of the host plant increased, by the presence of mycorrhizal infections. Ectotrophic mycorrhizas are abundant near the soil surface and it seems likely that they enable nutrient elements mineralized during the decay of litter to be efficiently re-absorbed into the plants.

Review of the distribution of ectotrophic mycorrhizas shows that they are strongly associated with conditions of mineral nutrient-stress. These include not only vegetation types such as heathland and sclerophyllous scrub in which the vegetation density is severely limited by the low mineral nutrient content of the habitat, but also mature temperatue and tropical forests in which mineral nutrients scarcity in the soil arises from the scale of nutrient absorption into the plant biomass.

In comparison with the roots of crop plants and species restricted to fertile soils, ectotrophic mycorrhizas appear to be long-lived, and the possibility may be considered that this attribute is itself a considerable advantage to a host plant growing under conditions of severe mineral nutrient stress. In competitive herbs adapted to exploit soils in which there is a large reservoir of available mineral nutrients, high rates of uptake are attained, firstly through the production of a very large absorptive surface composed of fine roots and root-hairs and, secondly, through continuous and rapid morphogenetic responses in root : shoot ratio and in the extension growth of individual parts of the root system. There is, however, continuous decay and replacement of roots, a process which is analogous to the rapid turnover of leaves which occurs simultaneously above ground in the same species. It is clear that this system of nutrient absorption, whilst effective in absorbing mineral nutrients, even from infertile soils, is achieved at the cost of a high expenditure of photosynthate and high rate of reinvestment of captured mineral nutrients. The main benefit of a mycorrhizal association to a host plant growing under conditions of mineral nutrient stress may therefore lie in the provision of an absorptive system which, because it remains functional over a long time and under varying conditions, allows exploitation of temporary periods of increased mineral nutrient availability, at relatively low cost to the synthetic resources of the plant.

Stress-tolerance and palatability

The possibility that the reduced stature and slow growth-rates of stress-tolerant plants may cause them to be particularly vulnerable to physical damage has been recognized by several authors (e.g. Whittaker, 1975). Plants growing under conditions of severe stress are likely to exhibit slow rates of recovery from defoliation by predators and during their long phase of establishment they will be particularly susceptible to the activity of herbivores. To species which exist for an extended period as small slow-growing plants, grazing, even by small invertebrates, presents a major threat to survival. It may be predicted, therefore, that many stress-tolerant plants will have experienced intensive natural selection for resistance to predation.

A comprehensive analysis of the frequency of unpalatable plants in stressed environments is not yet possible because of the shortage of reliable data and because it is often difficult to allow for the fact that palatability is a variable attribute depending upon the characteristics of plant and predator and the circumstances in which they interact (Tribe, 1950). Despite these complications, however, it is possible to refer to a large amount of circumstantial evidence (see Levin, 1971; Janzen, 1973; Feeny, 1975; Rhoades and Cates, 1976) which supports the hypothesis that in habitats subject to severe environmental stress there is a general decline in palatability.

In unproductive habitats, both physical and chemical deterrents to herbivory are conspicuous. These include the spines of many cacti and succulents, the hard or leathery texture of the foliage of many arid-zone shrubs and shade-plants, and the coarsely siliceous or needle-like leaves of many of the perennial herbs present in arid, arctic–alpine, or nutrient-deficient habitats. Evidence of low palatability and low rates of consumption by mammals is available for desert scrub communities (Chew and Chew, 1970) whilst the limited scope for herbivory in montane rain-forests has been commented upon by Leigh (1975) who pithily describes the coriaceous leaves of the dominant trees as 'built to last, and filled with poisons to keep off hungry insects.' In both temperate and tropical environments gastropods often provide a major hazard to survival, particularly at the seedling stage, and it is interesting to note that for both slugs (Cates and Orians, 1975) and snails (Grime *et al*., 1968) there is evidence that the palatability of vascular plants from unproductive or late-successional vegetation is lower than that of ruderal species.

With respect to both vertebrate and invertebrate herbivores circumstances have been described (e.g. Nicholson, Paterson, and Currie, 1970; Feeny, 1968, 1969, 1970; Rhoades and Cates, 1976) in which foliage is vulnerable to attack only during the phase of leaf expansion. It would seem quite possible, therefore, that the slow turnover of leaves in many stress-tolerators may contribute to the mechanisms resisting predation.

Examples of chemical defence against herbivores (e.g. Harris, 1960; Smith, 1966; Rhoades, 1976) are evident in the strongly aromatic or resinous compounds which, either as leaf constituents or when released into the atmosphere, deter predators in a range of herbaceous vegetation types including the chaparral of North America, the garigue of Southern Europe and the mulga of Central Australia. As several authors such as Levin (1971), Mooney (1972), and Whittaker (1975) have suggested, synthesis of chemical deterrents to predation or microbial attack, in some species, accounts for a considerable fraction of the photosynthate. The possibility may be recognized that 'defence expenditure' involving either physical or chemical mechanisms is a factor contributing to the low potential growth-rates of some stress-tolerant plants.

It is interesting to note that many lichens are not subject to severe predation. In this connection, it may be significant that the biochemistry of lichens is characterized by the production of a wide range of compounds (described collectively as the 'lichen substances') of no known metabolic function.

From a wide range of sources, including Davis (1928), Muller and Muller (1956), Naveh (1961), Landers (1962), and Del Moral and Muller (1969), there have been reports of phytotoxic effects of chemicals originating from the living parts or from the litter of plants adapted to stressed environments. Whilst these phenomena deserve consideration as evidence of allelopathic mechanisms (see page 142), the possibility may be recognized that such effects could arise from the release and persistence of compounds which have evolved primarily as a defence against predation.

RUDERALS

A definition of disturbance

Reference has been made in the preceding section to habitats in which the density of living and dead plant material remains low because production is severely restricted by environmental stress. It is quite clear, however, that low vegetation densities are not confined to unproductive habitats. Some of the world's most productive terrestrial environments, including many arable fields and pastures, are characterized by a rather sparse vegetation cover. Here the low densities arise from the fact that the vegetation is subject to partial or total destruction.

Although they are more obvious in particular habitats such as arable land, the effects of damage upon vegetation are ubiquitous. The amount of vegetation and the ratio of living to dead plant material in any habitat at any point in time depend upon the balance obtaining between the processes of production and destruction.

Even within one environment there may be considerable variety in the mechanisms which bring about the destruction of living or dead vegetation components. In addition to natural catastrophies (e.g. floods and windstorms) and the more drastic forms of human impact (e.g. ploughing, mowing, trampling, and burning), account must be taken of more subtle effects such as those due to climatic fluctuations and the activities of herbivores, decomposing organisms, and pathogens.

For the purpose of analysing the primary mechanism of vegetation, therefore, a term is required which encompasses this wide range of phenomena yet is capable of a simple definition. The term which will be used here is *disturbance*, which may be said to consist of *the mechanisms which limit the plant biomass by causing its partial or total destruction*.

Forms of disturbance differ with respect to their selectivity. Whilst the effects of herbivores and decomposing organisms tend to be restricted to the living or the dead material, respectively, certain phenomena, e.g. fire, may affect both components of the vegetation. In general, there is an inverse correlation between degree of selectivity and intensity of disturbance, a relationship which may be exemplified at one extreme by molecular discrimination between litter constituents by microbial decomposing organisms and at the other by the total vegetation destruction associated with phenomena such as severe soil erosion.

With respect to the intensity of damage experienced by the vegetation, a continuous range of plant habitats may be recognized. One end of the spectrum is represented by relatively undisturbed habitats such as mature temperate forests, in which the loss of plant material is a more or less continuous process associated with a comparatively low rate of predation and with senescence and the decomposition of litter (Odum, 1969). At the other extreme there are habitats such as arable fields in which, at frequent intervals, a large proportion of the plant biomass is summarily destroyed.

Among the forms of disturbance which affect living components, a distinction may be drawn between mechanisms which involve the immediate removal of plant structures from the habitat (e.g. grazing, mowing) and those in which plant material is killed but remains *in situ* (e.g. frost, drought, application of herbicides). In the latter, destruction proceeds in two stages. An initial rapid loss of solutes is followed by a rather longer phase in which the residue of plant structures is attacked by decomposing organisms.

Vegetation disturbance by climatic fluctuations

Some of the most important forms of severe disturbance leading to the development of ruderal vegetation are effects of climate. Since climatic factors such as low temperature and low rainfall have been associated already with the quite different phenomenon of stress (pages 23–25) it is necessary to draw a distinction between the circumstance in which the main effect of an unfavourable climate is merely to reduce productivity and that in which the influence is disruptive. Whether factors such as low rainfall result in disturbance or in stress depends* not so much upon their severity as upon the constancy of their occurrence. Vegetation disturbance by climate is prevalent where conditions encouraging the establishment (but not the uninterrupted growth) of competitors alternate abruptly with seasons imposing severe stress. In these circumstances neither competitors nor stress-tolerators can gain a secure advantage, and natural selection favours potentially fast-growing ephemeral plants. Few ecologists are likely to challenge the suggestion that marked fluctuations in water supply and temperature can act as agents of vegetation disturbance. Perhaps rather more controversial is the assertion (Al-Mufti and Grime, 1979) that in temperate woodlands on fertile soils the seasonal reduction in light intensity resulting from the expansion of a deciduous tree canopy functions primarily (through its association with shade-induced pathogenicity–see apge 139) as an agent of disturbance of the under-lying populations of herbs, shrubs, and tree seedlings. Here an important piece of evidence is the established fact (e.g. Scurfield, 1953; Al-Mufti *et al*., 1977) that in deciduous woodlands on very infertile soils no marked floristic changes or major disturbance coincides with canopy expansion, and the ground floras tend

*Both drought and temperature may cause stress *and* disturbance within the same habitat (pages 63–66).

to consist of stable communities of slow-growing evergreen herbs. This observation suggests that the onset of shading by deciduous trees has a disruptive effect on the herb layer only in systems which are sufficiently fertile to permit the establishment of shade-intolerant competitive or ruderal plants in the light phase during the early spring. On infertile soils, the conditions remain unfavourable to these plants throughout the year, and the effect of canopy expansion in the spring is merely to add the limiting effect of shade to that of mineral nutrient stress.

Adaptation to frequent and severe disturbance in various types of habitat

Just as the impact of stress upon vegetation changes according to its intensity (page 22), so also does that of disturbance. As described on page 13, in potentially productive enrironment,* low intensities of disturbance function as a modifier of competition by favouring those species which tend to maintain their competitive ability either by avoidance of damage or by rapid recovery from its effects. However, when we turn to productive habitats exposed to repeated and severe disturbance it is apparent that competitors are excluded and that a quite different strategy, that of the ruderal, prevails.

In order to recognize the basic characteristics of ruderal plants, the vegetation in five types of habitat subject to regular and severe disturbance will now be examined.

Ruderals of the sea-shore

Where sea coasts and inlets are composed of unstable sand or shingle, the effect of wave action and daily inundation is to prevent the establishment of vascular plants on the lower parts of the beach. However, at a position just above the normal upper limit of the tide—the so-called drift-line—a sparse vegetation commonly occurs. On the coasts of Britain, the most characteristic species of the drift-line are *Salsola kali*, *Cakile maritima*, and *Matricaria maritima*, all three of which are annual plants and are apparently resistant to salt-spray. Drift-line species are often rooted in organic debris, including the decaying remains of seaweed, and in this habitat they are frequently associated with potentially fast-growing annual plants such as *Galium aparine*, *Stellaria media*, and *Senecio vulgaris*, which are familiar in a variety of other disturbed habitats.

The drift-line vegetation is subject to frequent disturbance at high tides and during storms, and the colonizing species suffer high rates of mortality. An outstanding adaptive feature of the drift-line plants appears to be the ability of the survivors to grow and to produce seeds rapidly during the relatively short intervals between disturbances.

*Adaptation to low intensities of disturbance in unproductive habitats is considered on page 63.

Ruderals of marshland

In areas marginal to open water such as the sides of rivers, lakes, reservoirs, ponds, and ditches, considerable areas of bare mud and silt may become available for colonization when the water level falls during dry periods. These substrata are usually moist and rich in available mineral nutrients and often support extremely rapid plant growth (van Dobben, 1967). As in the case of the drift-line, the period available for growth may be relatively short and it is therefore no surprise to find that the species which exploit disturbed marshland are annuals of high potential growth-rate. This point may be illustrated by reference to Table 5 which includes estimates of the potential maximum relative growth-rates measured in populations of a number of annual species which are frequent colonists of bare mud and silt in the British Isles.

Arable weeds

In relation to the time-scale for the evolution of modern flowering plants, agriculture and the attendant expansion in the populations of arable weeds are comparatively recent events (Godwin, 1956). It seems reasonable to conclude therefore that, prior to the advent of agriculture, the majority of arable weeds had already evolved in habitats subject to more 'natural' forms of disturbance. This conclusion is supported by the fact that many arable weeds are of common occurrence in disturbed habitats in marshes and on the seashore.

Table 5. Estimates of the rates of dry matter production in the seedling phase in six ruderal plants compared to those attained by two crop plants grown under the same conditions (Grime & Hunt 1975). Measurements were conducted over the period 2–5 weeks after germination in a controlled environment (temperature: 20°C (day) 15°C night; daylength: 18 h; visible radiation: 38.0 W m^{-2}; rooting medium: sand + Hewitt nutrient solution). Rates of dry matter production are expressed as maximum (R_{max}) and mean ($\bar{R}$) relative growth rates and as a function of leaf area (E_{max}).

Species	Seed weight (mg)	R_{max} week^{-1}	$\bar{R}$ week^{-1}	E_{max} g m^{-2} week^{-1}
Ruderals				
Poa annua	0.26	2.7	1.7	120
Polygonum aviculare	1.5	1.4	1.4	134
Polygonum persicaria	2.1	1.3	1.3	177
Stellaria media	0.35	2.4	2.1	101
Veronica persica	0.52	1.7	1.3	75
Rorippa islandica	0.07	2.3	1.6	60
Crop plants				
Lycopersicon esculentum (Tomato)	3.1	1.6	1.0	64
Hordeum vulgare (Barley)	37.8	1.6	0.9	64

The majority of arable weeds are annuals of high potential growth-rate (Table 5) and under favourable conditions each plant is usually capable of producing a very large number of seeds (Salisbury, 1942), many of which tend to become buried and remain dormant in the soil for long periods. The significance of these and other characteristics of the reproductive biology of annual weeds is examined in Chapter 3.

Ruderals of trampled ground

Another group of habitats in which plants have evolved in relation to severe and persistent disturbance comprises the grasslands exploited by herds of wild or domesticated grazing animals. Particularly where damage by animals coincides with seasonal effects of drought, pastures may be composed almost exclusively of annual grasses, e.g. *Poa annua* and *Bromus mollis*, and annual legumes, e.g. *Trifolium dubium* and *Medicago lupulina*. Conditions favourable to ruderals arise in grassland not only from excessive trampling by the larger mammals including man, but also as a result of local effects such as uprooting of plants, scorching by urine, and burrowing and scraping by smaller animals. In farmland the most severe damage arises on trackways, and here again the most successful species are annual plants, many of which (e.g. *Poa annua* and *Polygonum aviculare*) are represented in trampled habitats by prostrate genotypes.

Desert annuals

In arid climates the vegetation is usually composed almost exclusively of stress-tolerant perennials (page 24). However, in areas where nonsaline and relatively fertile soils are exposed to a wet season, circumstances favourable to rapid growth may occur for a relatively short period. Under such conditions competitive species are excluded by drought, but annual plants may contribute significantly to the flora. Numerous studies of the ecology of desert annuals (e.g. Went, 1948, 1949; Koller, 1969) have drawn attention to the ability of these plants to persist as dormant seeds in the dry season and to germinate, grow, and to produce seeds extremely rapidly during periods in which moisture is temporarily available.

General features of ruderals

Life-cycle

It would appear that flowering plants adapted to persistent and severe disturbance have several features in common. The most consistent among these is the tendency for the life-cycle to be that of the annual or short-lived perennial, a specialization clearly adapted to exploit environments intermittently favourable for rapid plant growth. A related characteristic of many ruderals is the capacity for high rates of dry matter production (Baker, 1965; Grime and Hunt, 1975; Table 5), a feature which appears to facilitate rapid completion of the life-cycle and maximizes seed production.

In many ruderals flowering commences at a very early stage of development. The process of seed ripening may be extremely rapid and it is not uncommon in genera such as *Polygonum, Atriplex*, and *Chenopodium* for flowers and ripe seed to occur in the same inflorescence. Such features appear to be fully consistent with the habitat conditions experienced by ruderals; particularly where the effect of repeated disturbance is to cause a high rate of mortality it may be expected that natural selection will favour the early production and maturation of seeds.

Even in the absence of disturbance, ruderal plants are short-lived and, in the majority of annual species, seed production is followed immediately by the death of the parent. In this respect the ruderal differs consistently from both the competitor and the stress-tolerator. The significance of this difference is examined on page 47.

Response to stress

In preceding sections (pages 22 and 34) it has been concluded that differences in the rate and extent of morphogenetic responses to stress are of crucial importance in distinguishing the physiology of competitors and stress-tolerators. Experimental data are available which suggest that the stress-responses of ruderal plants are in certain respects quite different from those which are characteristic of either of the other two primary strategies. Before considering this evidence, it may be helpful to remember that two forms of stress, i.e. shortages of water and mineral nutrients, which ruderals, in common with other plants, experience in nature, may arise either from the direct impact of an unfavourable environment or may be due to the depletion of resources by neighbouring plants. There is no reason to suspect that a plant may differentiate in its response according to whether a shortage of water or mineral nutrients is plant-induced or not. This is a fact of considerable importance since, as explained in Table 7, the consequences of particular forms of response vary according to ecological context.

An informative method of studying the response of ruderal species to stress is to compare performance, and in particular seed production, of monospecific cultures sown at high and low densities. One of the first experiments of this type was that of Salisbury (1942) who compared capsule production in populations of a cornfield poppy (*Papaver argemone*) sown at high and low density on the same soil. The results of this experiment revealed that at the high sowing density the population showed a roughly normal distribution with respect to capsule number and included some plants with more than one hundred capsules. The most important fact to emerge from the experiment was that, although at the high sowing density mineral nutrient stress (perhaps associated with some mutual shading) caused severe reductions in vegetative development and seed production, *each plant produced at least one capsule*. More refined experiments (e.g. Hodgson and Blackman, 1956; Bunting, 1956; Hickman, 1975; Raynal and Bazzaz, 1975; van Andel and Vera, 1977; Foulds, 1978) which have been conducted on other species subsequently have confirmed the tendency of annual crops and weeds to sustain seed production, at however reduced a level (Plate 7), when subjected to severe stress.

One of the most interesting and thorough investigations of the stress physiology of a ruderal plant is that of Kingsbury *et al.* (1976) who examined the response of the annual halophyte *Lasthenia glabrata* to various periods of osmotic stress of the type normally experienced in the natural habitat of the species. Salt-water treatments applied at different stages of the life-cycle were found to induce 'a pulse of increased reproductive activity within a few weeks after stress began' and the authors concluded that 'osmotic stress from high salt concentrations causes a shift in the hormone balance to favour reproductive development over vegetative growth.' It was observed that prolonged exposure to salt stress conditions induced 'a high ratio of flowers to biomass, a relatively low vegetative yield, a higher probability of early senescence', and a life-cycle which was 'accelerated and condensed.'

Information relating to the life-cycles and stress-responses of ruderal plants is highly relevant to studies of the interactions between annual crops, and in a large number of experiments interactions between annual plants have been examined by sowing two species together in 1 : 1 mixtures and in various unequal proportions (e.g. DeWit, 1960, 1961). One of the most interesting features of the results of these experiments is the extent to which, even in experiments employing high total densities of seed, the species composition of the original seed mixture has remained the overriding determinant of the contribution to the total yield made by each of the two species. It seems reasonable to presume that this is due to the short-term and inconclusive nature of competition between ruderal plants. This in turn may be related in part to the limited capacity of the majority of annual plants for vegetative expansion, continuing resource capture, and monopolization of the environment. It seems likely also that an important contributory factor here is the tendency of ruderal plants, when impinging upon each other, to respond to the resulting stresses by growth-responses which maximize seed production at the expense of a rapid curtailment of vegetative development.

CONCLUSIONS

The theory of C-, S-, and R-selection

The information reviewed in this chapter suggests that, during the evolution of plants, the established phase of the life-cycle has experienced, in different habitats, three fundamentally different forms of natural selection. The first of these (C-selection) has involved selection for high competitive ability which depends upon plant characteristics which maximize the capture of resources in productive, relatively undisturbed conditions. The second (S-selection) has brought about reductions in both vegetative and reproductive vigour, adaptations which allow endurance of continuously unproductive environments. The third (R-selection) is associated with a short life-span and with high seed production and has evolved in severely disturbed but potentially productive environments.

We may also conclude that R-selected (ruderal), S-selected (stress-tolerant), and C-selected (competitive) plants each possess a distinct family of genetic characteristics and an attempt has been made in Table 6 to list some of these.

Three types of plant response to stress

It would appear that a crucial genetic difference between competitive, stress-tolerant, and ruderal plants concerns the form and extent of phenotypic response to stress. Many of the stresses which are a persistent feature of the environments of stress-tolerant plants are experienced, although less frequently and in different contexts, by competitive and ruderal plants. From the evidence reviewed in this chapter it would appear that competitive, stress-tolerant, and ruderal plants exhibit three quite distinct types of response to stress and we may conclude that such differences constitute one of the more fundamental criteria whereby the three strategies may be distinguished.

In Table 7 an attempt has been made, firstly to compare the circumstances in which competitive, stress-tolerant, and ruderal plants are normally exposed to major forms of stress, secondly to describe the stress response which appears to be characteristic of each strategy, and thirdly to predict the different consequences attending each type of response in different ecological situations.

It is concluded that the slow rate and relatively small extent of morphogenetic response and high physiological adaptability which is associated with endurance of protracted and severe stress in stress-tolerant plants is of low survival value in environments where stress is a prelude to either competitive exclusion or to disturbance by phenomena such as drought. Similarly, the rapid and highly plastic growth-responses to stress of competitive plants (tending to maximize vegetative growth) and of ruderals (tending to curtail vegetative growth and maximize seed production) are highly advantageous only in the specific circumstances associated, respectively, with competition and disturbance.

To summarize, therefore, we may conclude that the stress-response of the ruderal ensures the production of seeds, those of the competitor maximize the capture of resources, whilst those of the stress-tolerator allow the conservation of captured resources. This, of course, is an extremely simplified account of the immediate consequences of the three types of response. In the context of the entire life-history the three types of stress-response can be regarded as components of the rather different mechanisms whereby resource capture and fitness are maximized in different types of environment.

A most interesting implication concerns the difference in method of resource capture exhibited by the competitors and the stress-tolerators. As explained on page 13, the effect of the stress-responses of the competitors, when coupled with the rapid turnover of leaves and roots, is to bring about a continuous spatial re-arrangement of the absorptive surfaces which allows the plant to adjust to changes in the distribution of resources during the growing season. This type of growth response, although highly effective in resource capture, involves high rates of reinvestment of these resources and is clearly adapted to exploit productive but crowded environments where the effect of localized resource depletion by the rapidly-growing vegetation is to create within the habitat severe and continuously changing gradients in the distribution of light, water, and mineral nutrients. In contrast with the 'foraging' growth responses of the roots and shoots of the competitor, resource-capture in the stress-tolerator appears to be a more conservative activity primarily adapted to exploit temporal variation in the availability of resources in chronically unproductive habitats, i.e. the absorptive surfaces of the plant are associated with long-lasting physiologically-adaptive structures which, at least on an annual basis, tend to remain in the same location and to exploit temporary periods during which resources become available.

Three types of life-history

It has been noted already (page 44) that, in the majority of ruderals, seed production is followed by the death of the parent plant. This phenomenon impinges upon an important principle which has interested several evolutionary theorists (e.g. Cole, 1954; Williams, 1966; Stearns, 1976; Ricklefs, 1977) and which concerns the partitioning of captured resources between parent and offspring and the optimizing of life-histories by natural selection in various environments.

It is clear that when resources are expended upon reproduction at an early stage of the life-history there is an increased risk of parental mortality. However, in environments as uncertain as those of the ruderal, high rates of mortality are inevitable and the cost of a marginally-increased rate of parental fatality is outweighed by the benefit of high fecundity. As we might expect therefore the result of natural selection in most ruderals has been the development of early and 'lethal' (Harper, 1977) reproduction.

In comparison with the habitats of ruderals, the environments colonized by competitors and stress-tolerators are characterized by a low intensity of disturbance. This has two main effects: the first is to reduce the risks of mortality in long-lived plants, whilst the second is to limit drastically the opportunities for seedling establishment. A high risk of mortality to juvenile members of the plant population is characteristic of the environments of both the competitor and the stress-tolerator but, despite this point of similarity, the evolution of life-histories has taken a rather different course in the two types of habitat. Whereas in the competitors, reproduction occurs at a relatively early stage in the life-history and usually involves the expenditure each year of a considerable proportion of the captured resources, the stress-tolerators commence reproduction later and tend to show intermittent reproductive activity

Table 6. Some characteristics of competitive, stress-tolerant and ruderal plants.

	Competitive	Stress-tolerant	Ruderal
(i) Morphology			
1. Life-forms	Herbs, shrubs and trees	Lichens, herbs, shrubs and trees	Herbs
2. Morphology of shoot	High dense canopy of leaves. Extensive lateral spread above and below ground	Extremely wide range of growth forms	Small stature, limited lateral spread
3. Leaf form	Robust, often mesomorphic	Often small or leathery, or needle-like	Various, often mesomorphic
(ii) Life-history			
4. Longevity of established phase	Long or relatively short	Long—very long	Very short
5. Longevity of leaves and roots	Relatively short	Long	Short
6. Leaf phenology	Well-defined peaks of leaf production coinciding with period(s) of maximum potential productivity	Evergreens, with various patterns of leaf production	Short phase of leaf production in period of high potential productivity
7. Phenology of flowering	Flowers produced after (or, more rarely, before) periods of maximum potential productivity	No general relationship between time of flowering and season	Flowers produced early in the life-history
8. Frequency of flowering	Established plants usually flower each year	Intermittent flowering over a long life-history	High frequency of flowering
9. Proportion of annual production devoted to seeds	Small	Small	Large

10. Perennation	Dormant buds and seeds	Stress-tolerant leaves and roots	Dormant seeds
11. Regenerative* strategies	V, S, W, B_s	V, B_Υ	S, W, B_s
(iii) Physiology			
12. Maximum potential relative growth-rate	Rapid	Slow	Rapid
13. Response to stress	Rapid morphogenetic responses (root–shoot ratio, leaf area, root surface area) maximising vegetative growth	Morphogenetic responses slow and small in magnitude	Rapid curtailment of vegetative growth, diversion of resources into flowering
14. Photosynthesis and uptake of mineral nutrients	Strongly seasonal, coinciding with long continuous period of vegetative growth	Opportunistic, often uncoupled from vegetative growth	Opportunistic, coinciding with vegetative growth
15. Acclimation of photosynthesis, mineral nutrition and tissue hardiness to seasonal change in temperature light and moisture supply	Weakly developed	Strongly developed	Weakly developed
16. Storage of photosynthate mineral nutrients	Most photosynthate and mineral nutrients are rapidly incorporated into vegetative structure but a proportion is stored and forms the capital for expansion of growth in the following growing season	Storage systems in leaves, stems and/or roots	Confined to seeds
(iv) Miscellaneous			
17. Litter	Copious, often persistent	Sparse, sometimes persistent	Sparse not usually persistent
18. Palatability to unspecialized herbivores	Various	Low	Various, often high

*Key to regenerative strategies (considered in detail in Chapter 3): V – vegetative expansion, S – seasonal regeneration in vegetation gaps, W – numerous small wind-dispersed seeds or spores, B_s – persistent seed bank, B_Υ – persistent seedling bank.

Table 7. Morphogenetic responses to desiccation, shading, or mineral nutrient stress of competitive, stress-tolerant, and ruderal plants and their ecological consequences in three types of habitat.

Strategy	Response to stress	Consequences		
		Habitat 1*	Habitat 2†	Habitat 3‡
Competitive	Large and rapid changes in root:shoot ratio, leaf area, and root surface area	Tendency to sustain high rates of uptake of water and mineral nutrients to maintain dry-matter production under stress and to succeed in competition	Tendency to exhaust reserves of water and/or mineral nutrients both in rhizosphere and within the plant; etiolation in response to shade increases susceptibility to fungal attack	Failure rapidly to produce seeds reduces chance of rehabilitation after disturbance
Stress-tolerant	Changes in morphology slow and often small in magnitude	Overgrown by competitors	Conservative utilization of water, mineral nutrients, and photosynthate allows survival over long periods in which little dry-matter production is possible	
Ruderal	Rapid curtailment of vegetative growth and diversion of resources into seed production		Chronically low seed production fails to compensate for high rate of mortality	Rapid production of seeds ensures rehabilitation after disturbance

*In the early successional stages of productive, undisturbed habitats (stresses mainly plant induced and coinciding with competition).

†In either continuously unproductive habitats (stresses more or less constant and due to unfavourable climate and/or soil) or in the late stages of succession in productive habitats.

‡In severely disturbed, potentially productive habitats (stresses either a prelude to disturbance, e.g., moisture stress preceding drought fatalities or plant induced, between periods of disturbance).

over a long life-history. Analysis of this difference requires some reference to the regenerative strategies of plants (Chapter 3) and to the process of vegetation succession (Chapter 5). Here, therefore, only a preliminary explanation will be attempted.

In the crowded but productive environments colonized by competitive plants, seedling establishment is restricted because the habitat is occupied by a vigorously-expanding mass of established vegetation. Hence, although successful competitors accumulate resources at rates sufficient to sustain abundant seed production, there is little opportunity for the population to expand its immediate frontiers by seedling establishment. As explained in Chapter 3, this dilemma has evoked several types of evolutionary response in competitive herbs, shrubs, and trees. One is evident in the high incidence of vegetative expansion, a form of asexual regeneration which is viable in dense vegetation and is compatible with the maintenance of high competitive ability (see page 81). A second is the production annually of numerous wind-dispersed seeds, which facilitate the colonization of new habitats. This form of regeneration is particularly common among the perennial herbs, shrubs, and trees which appear in the early and intermediate stages of vegetation succession in fertile, undisturbed habitats. Seed production in these plants commences relatively early in the life-history and appears to be related to the fact that as vegetation development proceeds and resource depletion occurs the environment becomes progressively less hospitable to the competitor. Early and continuous expenditure on wind-dispersed seeds in many competitors may be interpreted, therefore, as an indication that, in common with many ruderals, competitors tend to lead a 'fugitive' (Hutchinson, 1951) existence.

Stress-tolerators occur in environments in which resources are severely restricted by absolute shortage or by the activity of the vegetation itself. In contrast to the populations of many competitors it seems likely that those of the stress-tolerators often remain in continuous occupation of the same habitat for many hundreds, perhaps thousands, of years. The biomass remains fairly constant and opportunities for reproduction may be exceedingly rare and dependent upon senescence or occasional damage to established members of the populations. In these circumstances, low parental mortality and a conservative but sustained reproductive effort can make a major contribution to the maintenance and expansion of population size. It is not surprising, therefore, that we should find that in stress-tolerators, such as *Pinus aristata, Pinus sembra,* and *Pseudotsuga menziesii* (Currey, 1965; Smith, 1970; Janzen, 1971), the onset of reproduction is delayed and flowering tends to occur only intermittently during a long life-history.

Reconciliation of C-, S-, and R-selection with the theory of r- and K-selection

The most generally accepted theory concerning strategies in the established phase to emerge from earlier studies with both animals and plants is the concept of r- and K-selection which was originally proposed by MacArthur and

Wilson (1967) and expanded by Pianka (1970). This theory has prompted many ecologists to recognize as opposite poles in the evolutionary spectrum two types of organism. The first, said to be K-selected, consists of organisms in which the life expectancy of the individual is long and the proportion of the energy and other captured resources which is devoted to reproduction is small. The second, or r-selected, type is made up of organisms with a short life-expectancy and large reproductive effort. It is now widely accepted that the majority of organisms fall between the extremes of r- and K-selection and the results of studies such as those of Gadgil and Solbrig (1972), Abrahamson and Gadgil (1973), and McNaughton (1975) suggest that genetic variation may cause populations of the same species to occupy different positions along the r–K continuum.

Figure 11 attempts to reconcile the concept of the three primary plant strategies with that of r- and K-selection. It is suggested that the ruderal and stress-tolerant strategies correspond, respectively, to the extremes of r- and K-selection and that competitors occupy an intermediate position. This relationship is consistent with the model proposed later (page 150) to explain the involvement of the three strategies in the process of secondary vegetation succession.

The most substantial way in which the three-strategy model differs from that of the r–K continuum lies in recognition of stress-tolerance as a distinct strategy evolved in intrinsically unproductive habitats or under conditions of extreme resource depletion induced by the vegetation itself. Inspection of Figure 11 suggests that there are two critical points (i), (ii) along the r–K continuum. At (i), the intensity of disturbance becomes insufficient to prevent the exclusion of ruderals by competitors, whilst at (ii) the level of supply of resources is

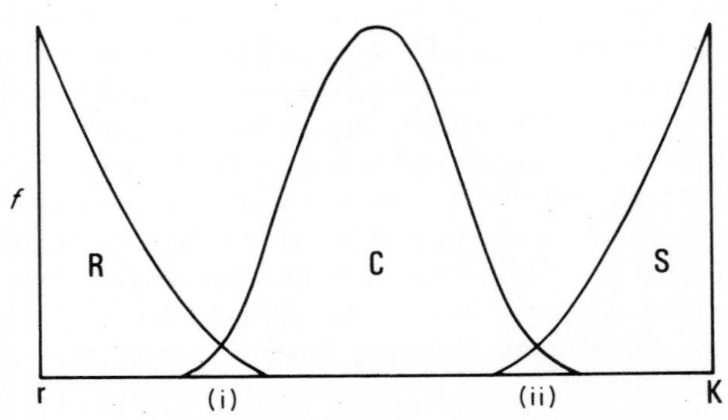

Figure 11 Diagram describing the frequency (*f*) of ruderal (R), competitive (C), and stress-tolerant (S) strategies along the r–K continuum. An explanation of the significance of critical points (i) and (ii) is included in the text (Grime, 1977). (Reproduced from *American Naturalist*, **111**, by permission of the University of Chicago Press. © 1977. The University of Chicago Press.)

depleted below the level required to sustain the high rates of reinvestment of captured resources characteristic of competitors and selection begins to favour the more conservative physiologies of the stress-tolerators.

Reservations have been expressed concerning the general validity of the theory of r- and K-selection by various authors such as Wilbur *et al.* (1974), Grubb (1976), Swingland (1977), and Moore (1978). It would seem, however, that most objections apply not to the theory itself but to the suites of characters which later authors (Pianka, 1970; Southwood *et al.*, 1974; Southwood, 1976) have attributed to r- and K-organisms. Many of the inconsistencies which become apparent when these extensions of MacArthur and Wilson's theory are applied may be resolved first by recognizing the phenomenon of uncoupling between juvenile and mature phases of life-cycles (see page 1 and Chapter 3) and by observing the distinction between C- and S-selection.

It is desirable that any general model of strategies in the established phase should describe also the secondary strategies which have evolved in the various conditions which fall between those associated with the primary strategies. In this respect the scheme represented in Figure 11 is both incomplete and misleading. This is because the linear arrangement does not correspond to the full range of conditions to which strategies of the established phase may be adapted. In order to accommodate the range of equilibria which may exist between stress, disturbance, and competition it is necessary to arrange the primary strategies within the triangular model considered in the next chapter.

Analogous strategies in fungi and in animals

In view of the evidence suggesting that in the established phase autotrophic plants conform to three basic strategies and to various compromises between them, it is interesting to consider whether the same pattern obtains in heterotrophic organisms.

In fungi, there is strong evidence of the existence of strategies corresponding to those recognized in green plants. 'Ruderal' life-styles are particularly characteristic of the *Mucorales* in which most species are ephemeral colonists of organic substrates. These fungi grow exceedingly rapidly and exploit the initial abundance of sugars, but as the supply of soluble carbohydrate declines they cease mycelial growth and sporulate profusely. 'Competitive' characteristics are evident in fungi such as *Serpula lachrymans* and *Armillaria mellea* which are responsible for long-term infections of timber and produce a consolidated mycelium which may extend rapidly through the production of rhizomorphs. Examples of 'stress-tolerant' fungi appear to include the slow-growing basidiomycetes which form the terminal stages of fungal succession on decaying matter and the various fungi which occur in lichens and ectotrophic mycorrhizas. All of the 'stress-tolerant' fungi are characterised by slow-growing, relatively persistent mycelia and low reproductive effort.

The majority of the attempts which have been made to recognize strategies in animals have adhered closely to the concept of the r–K continuum. However,

Wilbur *et al.* (1974), Whittaker (1975), Nichols *et al.* (1976), and Southwood (1977a) have recognized the need for an extended theoretical framework and have presented information and ideas consistent with the concept of three primary strategies.

In the animal kingdom, 'ruderal' life-histories occur in both herbivores and carnivores. They are particularly common in insects (e.g. aphids, blow-flies) where, as in ruderal plants, populations expand and contract rapidly in response to conditions of temporary abundance in food supply (e.g. Brown, 1962; Hughes, 1970; Southwood, 1977b). As pointed out by Southwood *et al.* (1974), similar characteristics are apparent in short-lived birds including not only grain-eaters, such as the zebra finch (*Taeniopygia castanotis*) and the budgerigar (*Melopsittacus undulatus*), but also the insectivorous bluetit (*Parus caeruleus*).

Earlier it has been pointed out that a characteristic of competitive plants is the possession of mechanisms of phenotypic response which maximize the capture of resources, and it has been argued that these mechanisms are advantageous only in environments in which high rates of capture of energy, water, and mineral nutrients can be sustained. It has been suggested (Grime, 1977) that the same principle applies in animals and that a direct analogy can be drawn between the 'foraging' responses of competitive plants (page 21) and the liberal expenditure of energy and other captured resources which characterizes the very active methods of food-gathering (e.g. Newton, 1964) observed in the species of mammals, birds, and fish which depend for their survival upon continuous access to high rates of food supply. Support for this hypothesis is evident in the theoretical analysis of Norberg (1977) who concludes that 'the most energy-consuming, but also most efficient, search methods should be employed by a predator at the highest prey densities.'

It is interesting to note that Norberg also concludes that 'when prey density decreases a predator should shift to progressively less energy-consuming search methods although these are connected with low search efficiency.' This same principle has been stated with specific relevance to thermally-inhospitable environments by Morhardt and Gates (1974) as follows: 'Regulation of energy exchange may require a considerable portion of an animal's time and metabolic energy reserves and, in energetically severe habitats, may seriously limit activities which are necessary for the normal biology of the animal, such as seeking food and defending territory.'

From these theoretical studies, and also by analogy with plant strategies, therefore, we may predict that in extreme habitats where food is scarce, the level of food capture may be too low and the risks of environmental stress (e.g. desiccation, hypothermia) and predation too great to permit domination of the fauna by species which indulge in very active methods of food gathering. Studies of certain of the animals which occur in habitats such as deserts (Chew and Butterworth 1964; French *et al.*, 1966; Pianka, 1966; Tinkle, 1969), coral reefs (Odum and Odum, 1955; Muscatine and Cernichiari, 1969; Goreau and Yonge, 1971), and tropical forests (Montgomery and Sunquist, 1974) indicate that, in species adapted to low rates of food supply, strategies of active food gathering give way to feeding mechanisms which are more conservative and appear to be analogous to the assimilatory specializations observed in stress-tolerant plants. Moreover, the similarities between plants and animals exploiting conditions of low resource-availability extend to the life-histories and reproductive biology. Some of the most remarkable parallels with stress-tolerant plants are to be found in the long-life histories and exceedingly delayed and intermittent reproduction in reptiles such as the sphenodon (*Sphenodon punctatum*), and the Aldabran giant tortoise (*Geochelone gigantea*) and birds such as the wandering albatross (*Diomedea exulans*). As Nichols *et al.* (1976) have pointed out, there is a scattered but extensive body of evidence which suggests that, in environments with low and uncertain rates of food supply, there is a tendency for animal life-histories to be extended and for reproductive activity to be capable of suspension during exceptionally unfavourable years. In desert environments in North America, for example, it is not unusual, during years of very low rainfall and reduced plant production, for reproduction to be completely inhibited in rodents belonging to genera such as *Perognathus* and *Dipodomys* (French *et al.*, 1966, 1967, 1974). Suspension of reproduction also occurs in various types of Australian animals of desert environments including birds (Keast, 1959) and frogs (Bragg, 1945; Main *et al.*, 1959; Bentley, 1966), and a similar phenomenon has been recorded in lizards (Turner *et al.*, 1973) and in waterfowl of unpredictable habitats (Frith, 1973).

Further parallels between plant and animal strategies are apparent from several studies in which measurements have been made of the effect of food shortage on the longevity and reproductive activity of invertebrate animals. Of particular relevance are the comparative investigations which have been conducted upon freshwater triclads (Reynoldson, 1961, 1968; Calow, 1977) and gastropods (Grime and Blythe, 1968) which show differences in response which closely parallel those distinguishing ruderal and stress-tolerant plants. In the studies of Calow and Woolhead (1977), for example, it has been shown that, in the annual triclad *Dendrocoelum lacteum*, the response to food shortage is to sustain reproductive effort with the result that increased rates of mortalities are experienced in the adult population. In marked contrast, under the same conditions, the perennial species *Dugesia lugubris* responds promptly with a cessation of reproduction, a mechanism which, as in the case of competitive and stress-tolerant plants, appears to safeguard the survival of the established population.

Chapter 2

Secondary strategies in the established phase

INTRODUCTION

The environments which, in Chapter 1, have been associated with the occurrence of competitors, stress-tolerators, and ruderals form only part of the spectrum of habitats available to plants. It seems reasonable to suppose, therefore, that in addition to the three extremes of evolutionary specialization there will be various secondary strategies which have evolved in habitats experiencing intermediate intensities of competition, stress, and disturbance.

The model drawn in Figure 12 illustrates the conditions in which various types of secondary strategies may be expected to occur. The model consists of an equilateral triangle in which variation in the relative importance of competition, stress, and disturbance as determinants of the vegetation is indicated by three sets of contours. At their respective corners of the triangle, competitors, stress-tolerators, and ruderals become the exclusive constituents of the vegetation and the remaining areas of the triangle correspond to the various equilibria which are possible between competition, stress, and disturbance. Four main types of secondary strategy are proposed. These consist of:

1. *competitive ruderals* (C–R)—adapted to circumstances in which there is a low impact of stress and competition is restricted to a moderate intensity by disturbance;
2. *stress-tolerant ruderals* (S–R)—adapted to lightly-disturbed, unproductive habitats;
3. *stress-tolerant competitors* (C–S)—adapted to relatively undisturbed conditions experiencing moderate intensities of stress;
4. *'C–S–R strategists'*—adapted to habitats in which the level of competition is restricted by moderate intensities of both stress and disturbance.

From both field and laboratory investigations, there is evidence confirming the existence of these strategies and it would appear that the criteria which have

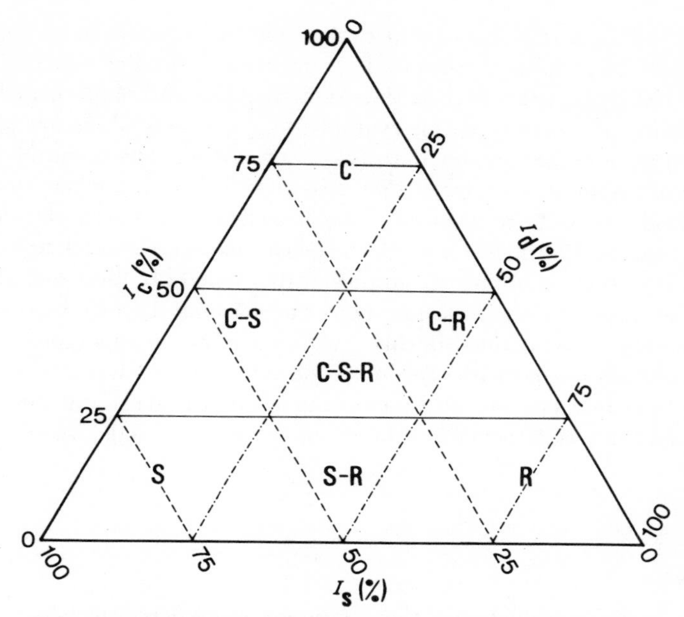

Figure 12 Model describing the various equilibria between competition, stress, and disturbance in vegetation and the location of primary and secondary strategies. I_c, relative importance of competition (——); I_s, relative importance of stress (– – –); I_d, relative importance of disturbance (–·–·–). A key to the symbols for the strategies is included in the text. (Reproduced from *American Naturalist*, **111**, by permission of the University of Chicago Press. ©1977. The University of Chicago Press.)

been used to define the primary strategies (Table 6) also provide a basis for recognition of the secondary strategies. Until comparative studies have been conducted on the ecology, life-histories, and physiology of a wider range of plants, a comprehensive account of the four secondary strategies cannot be attempted. However, mainly by reference to herbaceous plants of temperate environments, an attempt can be made to illustrate some of their key characteristics. Although, in the descriptions which follow, attention is confined to the established phase of the life-cycle, it is worth noting that certain of the secondary strategies usually occur in combination with particular regenerative strategies. An assessment of the significance of certain associations between secondary strategies and regenerative strategies is attempted at the end of Chapter 3.

COMPETITIVE-RUDERALS

The competitive-ruderals occur in habitats of high productivity in which dominance of the vegetation by competitors is prevented by disturbance. In comparison with habitats populated exclusively by ruderals, *sensu stricto*, the sites colonized by competitive-ruderals experience a smaller effect of disturbance. The reduced impact may be due to the fact that although disturbance is severe it occurs infrequently as, for example, in grasslands which are ploughed and re-sown at intervals greater than two years. Conditions favourable to the persistence of competitive-ruderals may also arise in habitats where damage to the vegetation occurs once annually and is sufficient to check the vigour of competitive species but does not reach the degree of severity necessary to eliminate them from the vegetation. Examples of this type of habitat include fertile meadows and productive grasslands subject to seasonal damage by drought or grazing animals. Also included in this category are the various types of vegetation which are associated with seasonal flood damage, silt deposition, and soil erosion on river terraces and at the margins of ponds, lakes, and ditches.

The competitive-ruderals fall into three classes: annuals, biennials, and perennials.

Annual herbs

The European flora includes many familiar annual plants which may be classified as competitive-ruderals. At river margins and on other types of disturbed fertile ground extensive populations of the annuals *Galium aparine* and *Impatiens glandulifera* occur. The leaves and stems of *Galium aparine* are covered with small hooks which allow the species to scramble to a considerable height over the shoots of perennial herbs and shrubs (Plate 4). In contrast, *Impatiens glandulifera* often occurs in extensive stands in which each individual is composed of an erect unbranched stem bearing several whorls of large leaves (Plate 9).

In North America, competitive ruderals may be recognized among the large summer annuals (e.g. *Ambrosia artemisiifolia*, *Polygonum pensylvanicum* and *Abutilon theophrasti*) which colonize abandoned arable fields. Flowering in these plants is preceded by a comparatively long vegetative phase (Raynal and Bazzaz, 1975). At maturity, the shoots may be several metres in height and where many individuals have established in close proximity a dense leaf canopy may be developed.

In these potentially large summer annuals, the degree of dominance (see page 124) depends not only upon the density of individuals but upon the duration and vigour of the vegetative phase, termination of which is usually determined by the induction of flowering by increasing daylength (Ray and Alexander, 1966; Cumming, 1969). It is clear that in the life-cycles of these species there is a delicate balance between the initial competitive phase and the later reproductive phase. Optimization of resource capture and seed production in the two phases of the life history results in some ecotypic variation, particularly with respect to latitudinal races of these plants; this has been considered with respect to *Chenopodium rubrum* by Cook (1976). We may also anticipate that the response to stress in these relatively long-lived annuals is likely to vary according to the stage of development, but little work appears to have been carried out on this subject.

Grasses also provide species which may be described as competitive-ruderals. Examples are *Hordeum murinum, Bromus sterilis,* and *Lolium multiflorum*, all species which are of common occurrence in grasslands in which the perennial component is subject to severe seasonal damage by droughting, grazing, and trampling. As in all the other species cited under this heading, each is capable of rapid dry matter production (Table 8) and under favourable conditions may reach a large size at maturity. It is interesting to note from Table 8 that there is some variation between these annuals with respect to the mechanism by which large size may be achieved during the vegetative phase. In certain species, e.g. *Polygonum convolvulus*, dry matter production depends upon the combination of a large seed reserve with a modest relative growth rate. In *Chenopodium rubrum*, however, the seed is small and large size may be attained only by a sustained period of rapid photosynthesis.

Many of the species employed in agriculture including *Hordeum vulgare* (barley), *Secale cereale* (rye), *Fagopyron esculentum* (buckwheat), *Zea mays* (maize), *Helianthus annuus* (sunflower), and *Panicum mileaceum* (millet) have characteristics (annual life cycle, potentially rapid relative-growth rates, capacity to produce a high leaf area index) which allow them to be classified as competitive-ruderals.

Biennial herbs

A characteristic component of many types of productive, intermittently-disturbed vegetation is made up of large biennial herbs. In temperate regions one

Table 8. Estimates of the rates of dry matter production in the seedling phase in ten annual competitive-ruderals (Grime & Hunt 1975). Measurements were conducted over the period 2–5 weeks after germination in a controlled environment (temperature: 20°C (day) 15°C (night); daylength: 18 h; visible radiation: 38.0 W m^{-2}; rooting medium: sand + Hewitt solution). Rates of dry matter production are expressed as maximum (R_{max}) and mean ($\bar{R}$) relative growth rates and as a function of leaf area (E_{max}). The seed weights quoted refer to the seed samples used.

Species	Seed weight (mg)	R_{max} $week^{-1}$	$\bar{R}$ $week^{-1}$	E_{max} $g\ m^{-2}$ $week^{-1}$
Bromus sterilis	8.4	2.3	1.4	134
Chenopodium rubrum	0.09	2.0		116
Galium aparine	7.3	1.5	1.2	96
Helianthus annuus	66.0	0.8	0.8	91
Hordeum murinum	6.6	1.8	1.2	102
Hordeum vulgare	37.8	1.6	0.9	64
Lolium multiflorum	2.5	2.0	1.3	100
Lycopersicon esculentum	3.1	1.6	1.0	93
Polygonum convolvulus	1.3	1.9	1.4	68
Zea mays	273.0	0.9	0.9	107

family, the *Umbelliferae*, provides a number of good examples including such familiar European species as *Anthriscus sylvestris, Heracleum sphondylium, Angelica sylvestris*, and *Conium maculatum*. Another family which contributes to this group is the *Compositae* which includes several large biennial species, e.g. *Cirsium vulgare, Carduus nutans*, and *Arctium* spp., which are of frequent occurrence in disturbed vegetation in many parts of the world.

The habitats of these large biennials resemble closely those exploited by the annuals described under the previous heading and it is not unusual for species of both groups to grow together. An obvious point of difference from the annuals lies in the fact that the life-history usually extends over two growing seasons. During the first, a rosette of leaves is formed and photosynthate is accumulated in a swollen rootstock. In the second season the storage organ forms the capital for expansion of flowering shoots which under favourable circumstances may become large structures bearing a considerable number of seeds.

In terms of the established strategy therefore it seems reasonable to regard these biennials as close relatives of the large annuals. The main difference is in the more formalized separation of the vegetative and reproductive phases of the life-history. A further distinction, which identifies the biennials as less ruderal in character, is the tendency under unfavourable conditions for the vegetative phase to become extended over several years. A good example of this phenomenon is provided by the study of Werner (1975) in which it was shown that in a North American population of *Dipsacus fullonum* there was a high probability that flowering would be delayed in circumstances where the diameter of the rosette failed to achieve a diameter of 30 cm by the end of the first season.

Ruderal-perennial herbs

A third group which may be recognized within the competitive-ruderals consists of a variety of plants which may be conveniently described as ruderal-perennials. These plants often occur as seedlings or small plants during the initial colonization of bare ground but they are most abundant in circumstances in which the impact of disturbance is less immediate or catastrophic, such as those which occur during the second and third year after colonization of bare soil in abandoned arable fields, derelict gardens, and construction sites. Species which belong in this category include forbs, e.g. *Ranunculus repens*, *Achillea millefolium, Cirsium arvense, Tussilago farfara*, and *Trifolium repens*, and grasses, e.g. *Poa trivialis, Poa pratensis, Agrostis stolonifera, Agropyron repens*, and *Holcus lanatus*.

The majority of these plants are strongly rhizomatous or stoloniferous and show a capacity for rapid vegetative spread. Life-history studies (Myerscough and Whitehead, 1967; Ogden 1974; Sarukhan, 1971; Bostock, 1976; Watt, 1976) of certain of these species have confirmed that, once established in fertile conditions, there is a rapid lateral extension of vegetative shoots either through the soil (e.g. *Tussilago farfara*) or over the ground surface (e.g. *Ranunculus repens*). In many of the ruderal perennials the life-span of an individual shoot is usually less than one year and this characteristic, coupled with the vigorous and often complex proliferation and fragmentation of rhizomes or stolons, brings about a new spatial distribution of shoots in each successive growing season. It is not surprising, therefore, to find that where the effect of seasonal disturbance is to produce a discontinuous vegetation cover, species of this type are efficient colonizers of temporary gaps.

Where the process of colonization of fertile sites by herbaceous vegetation is allowed to proceed undisturbed for several years the ruderal perennials are progressively excluded by the presence of perennial herbs with taller and more consolidated growth forms, which are capable of producing a dense cover of foliage and plant litter in which there are few gaps. Scrutiny of the vegetation types in which ruderal perennials are a persistent component (e.g. heavily-disturbed pastures, meadows, and marshes) suggests that these plants are dependent upon conditions in which the intensity of damage sustained each year by the vegetation is sufficient to debilitate more strongly competitive perennial herbs.

The general characteristics of competitive-ruderals

From the preceding review of the characteristics of the annuals, biennials, and perennials included within the competitive-ruderals it is possible to recognize several points of difference from the ruderals *sensu stricto*. In general, competitive ruderals exhibit a longer period of vegetative growth, and a considerable biomass may be attained before the onset of flowering. A consequence of the delay in seed production is that populations of competitive ruderals are more susceptible than those of ephemerals to habitat disturbance during the growing season. A further distinction lies in the ability of some of the competitive-ruderals to exploit environments already occupied by perennial species. However, it is clear that co-existence with vigorously-growing perennial species occurs only under specific conditions. Phenological studies by Al-Mufti *et al.* (1977), for example, illustrate that in Britain the competitive-ruderals *Galium aparine* and *Poa trivialis* persist in vegetation dominated by larger perennials (either herbs or trees) by exploiting periods in the year when the potential impact of the dominant species is at a minimum. An example of this phenomenon is illustrated in Figure 13 which describes the relationship between the shoot phenology of *Poa trivialis* and the seasonal change in standing crop and tree litter on an alluvial terrace which in summer is shaded by a dense canopy of deciduous trees. From this figure, it is apparent that *Poa trivialis* exhibits a truncated phenology in which rapid growth and flowering occurs in the short period in the spring preceding full expansion of both tree and herbaceous canopies and coinciding with the annual minimum in density of tree litter. In roadsides, meadows, and river banks a similar phenology occurs in the biennial *Anthriscus sylvestris* and in the annual *Impatiens glandulifera*, in each of which leaf expansion usually occurs in advance of that of the neighbouring shoots of established perennial herbs.

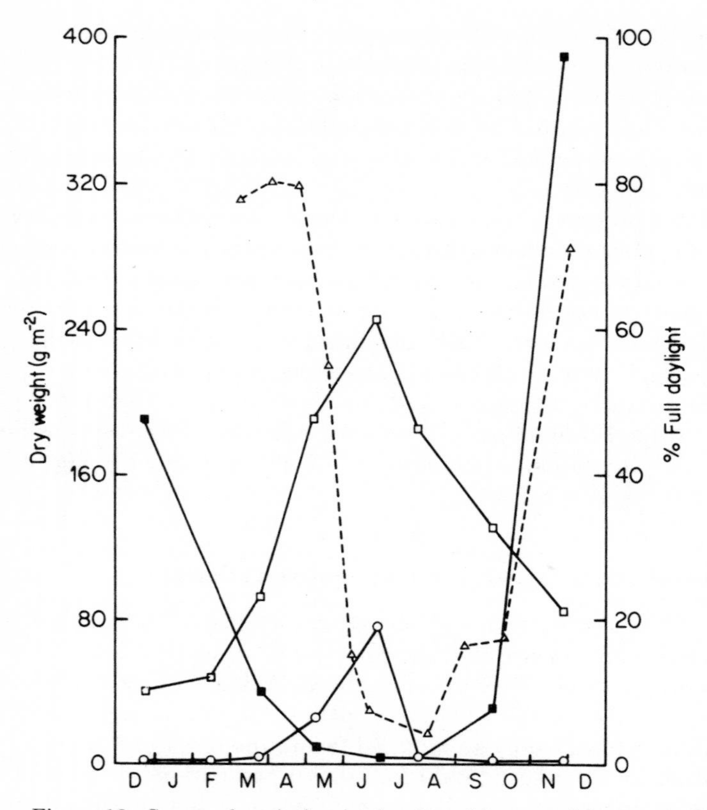

Figure 13 Seasonal variation in the shoot biomass of the perennial grass *Poa trivialis* (o–o) in a deciduous woodland in Northern England, and associated changes in the total biomass of the herb layer (□–□) and in the density of tree litter (■–■). Light intensity reaching the herb layer (△–△) is expressed as a percentage of that measured simultaneously at an adjacent unshaded site (Al-Mufti *et al*., 1977).

Although the annuals, biennials, and perennials are distinguishable in terms of life-history and reproduction there can be little doubt of the strong ecological affinities between the three classes of competitive-ruderals. There are many vegetation types in which representatives of two or all three of the classes occur together. Reference has been made already to the frequent association of the annual species *Galium aparine* with the short-lived perennial *Poa trivialis*. In damp meadows and marshes it is not unusual to find biennials such as *Heracleum sphondylium* and *Angelica sylvestris* in close proximity to perennial species such as *Ranunculus repens* and *Agrostis stolonifera*, whilst in semi-derelict habitats in urban areas all three classes of competitive ruderals are often represented.

In view of this evidence of close similarities in field distribution it is pertinent to consider why the different life-histories exhibited by the three classes of competitive ruderals do not give rise to greater disparities in ecology. In particular it is necessary to explain why species totally dependent upon seedling establishment for maintenance of their populations (the annuals and biennials) frequently co-exist with perennial ruderals in which the annual replacement of individuals is mainly by vegetative reproduction. However, as we shall see in Chapter 3, the regenerative strategies of the annuals and biennials are not as dissimilar from the perennials as they appear on first inspection.

STRESS-TOLERANT RUDERALS

Stress-tolerant ruderals occur in habitats in which the character of the vegetation is determined by the coincidence of moderate intensities of stress and disturbance. The stress-tolerant ruderals resemble the ruderals *sensu stricto* in that opportunities for growth and reproduction are usually restricted to relatively short periods. However, a distinguishing characteristic of the stress-tolerant ruderals is that they occupy habitats in which stress conditions are experienced during the period of growth. The severity of stress often varies from one growing season to the next, but it is usually sufficient to restrict annual production to well below that achieved in habitats dominated by ruderals or competitive-ruderals and in exceptional years both vegetative growth and reproduction may be very severely restricted.

Two types of plants are strongly represented among the stress-tolerant ruderals. The first consists of small herbs whilst the second is made up of bryophytes.

Herbaceous plants

The herbaceous plants which may be classified as stress-tolerant ruderals consist of a rather heterogeneous assemblage of plants drawn from widely different habitats encompassing the complete range of world climatic zones. Two main groups may be recognized. The first consists of small annuals and short-lived perennials whilst the second is composed of small geophytes.

Small annuals and short-lived perennials

Terrestrial habitats which experience severe levels of both stress and disturbance tend to be devoid of vegetation; the effect of low productivity and disturbance is such that even short-lived species are unable to complete their life-cycles. However, as we approach conditions in which the effects of stress and disturbance are slightly less severe, environments are encountered in which certain specialized plants can survive. Among the latter are a variety of small annual plants. These include the tiny arctic annual *Koenigia islandica*, the diminutive desert ephemerals which emerge following rain-showers in some deserts (Went, 1948, 1949, 1955), and the wide variety of annuals which exploit shallow unproductive soils in temperate regions. By far the largest contribution to this last group is that composed of the various types of small winter annuals, many of which have been the subjects of intensive studies in Europe (Ratcliffe, 1961; Newman, 1963; Pemadesa and Lovell, 1974), North America (Tevis, 1958; Beatley, 1967), and Australia (Mott, 1972).

In temperate regions these plants are restricted to shallow or sandy soils which experience desiccation during the summer or to local habitats such as ant-hills (Grubb, Green, and Merrifield, 1969; King, 1977a, b, c) where the vigour of perennial herbs is restricted by animal activity. Growth and flowering occur during the cool wet season and the seeds remain ungerminated throughout the dry summer period. In comparison with summer annuals or the winter annuals which occur on deep fertile soils (e.g. *Galium aparine*) the stress-tolerant winter annuals are small in stature (Plate 10), have lower potential relative growth-rates, and produce smaller seeds. Such characteristics are consistent with the fact that these plants occur in habitats in which production is often severely restricted by mineral nutrient deficiencies and grow at times in the year when photosynthesis is restricted by low temperature.

It is not uncommon, especially at more northern latitudes, for the life-cycles of certain winter-annuals, e.g. *Erigeron acer, Acinos arvensis, Linum catharticum*, to be extended over two years while some of the rather less ephemeral monocarpic species of disturbed unproductive habitats, e.g. *Carlina vulgaris, Gentianella amarella*, have life-histories which frequently extend for longer periods. Like the small winter annuals, these biennials and short-lived perennials are particularly common in unproductive habitats in which the perennial component is subject to local damage by drought or by biotic factors, e.g. hoof-marks and rabbit-scrapes. Again, in common with the winter annuals, the short-lived perennials show considerable plasticity in growth form and under unusually favourable conditions numerous seeds may be produced.

Small geophytes

Another distinctive component of disturbed but unproductive vegetation is composed of what Noy-Meir (1973) has described as the 'perennial ephemeroids'. Many of these plants resemble the small winter annuals in that they are abundant on shallow unproductive soils in regions such as Southern Europe where herbaceous vegetation is subject to severe desiccation during the summer. Growth in these plants tends to be confined to the moist, cool season (Figure 14) and survival of the dry season is in the form of an underground storage organ such as a bulb (e.g. *Scilla verna*), tuber (e.g. *Orchis mascula*), or rhizome (*Anemone nemorosa*, *Primula veris*).

It is evident that in several respects the geophytes resemble the group of annual and biennial species considered under the last heading. Traits common to both types include small stature, rather slow relative growth-rates, and seeds which vary in size from small to minute. All three of the characteristics appear to be related to the low productivity of the habitat. A possible point of

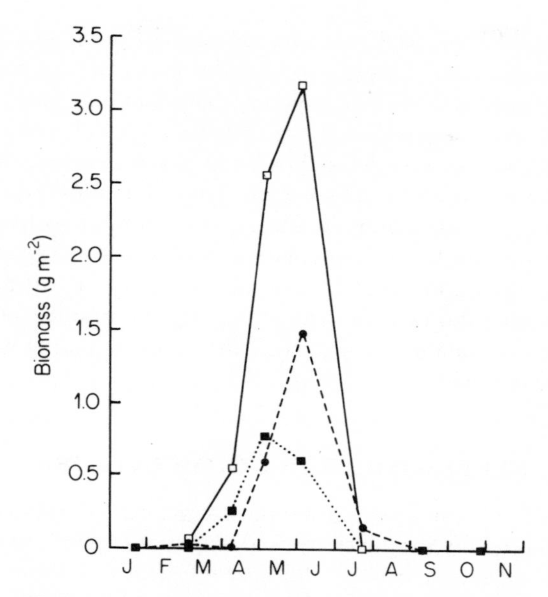

Figure 14 Seasonal change in the shoot biomass in three small vernal geophytes, *Anemone nemorosa* (□—□), *Primula veris* (●—●), and *Orchis mascula* (■—■), present in a species-rich calcareous pasture in Northern England (Furness, 1978).

difference between the annuals, short-lived perennials, and geophytes relates to the year-to-year fluctuations in population density. Following years of exceptional disturbance by factors such as drought, trampling, or fire, annual and biennial species often show very rapid expansion in numbers (e.g. Lloyd, 1968). Populations of geophytes appear to be relatively more stable, although in some species, including many orchids in which the juvenile plants are saprophytic and remain below ground, it is necessary to conduct long-term population studies (e.g. Tamm, 1972; Wells, 1967) in order to obtain accurate estimations of population size.

Bryophytes

The least conspicuous of the stress-tolerant ruderals are mosses and liverworts. This is not to suggest that all bryophytes conform to this strategy. Many mosses, e.g. *Rhacomitrium lanuginosum, Dicranum fuscescens*, occur in extremely unproductive habitats and studies of their ecology, life-history, and physiology (Tallis, 1958, 1959, 1964; Oechel and Collins, 1973; Hicklenton and Oechel, 1976) have shown that they have strong affinities with lichens whilst others, e.g. *Funaria hygrometrica* (Hoffman, 1966), are ephemeral colonists of disturbed habitats. However, in addition to these stress-tolerators and ruderals there are many bryophytes, e.g. *Brachythecium rutabulum,* which are associated with mesic conditions and exhibit marked seasonal changes in biomass (Al-Mufti *et al.*, 1977; Furness *et al.*, 1978). This group includes many of the mosses and liverworts which, in temperate regions, colonize bare soil and plant litter beneath productive herbaceous vegetation (Plate 29). In such habitats, the bryophyte component usually shows well-defined peaks of growth in the moist, cool conditions of early spring and autumn (see Figure 53). The most likely explanation for the decline in biomass observed during the summer would appear to be desiccation, although under certain conditions effects of shade and high temperatures also may be implicated. The results of laboratory studies (see Figure 54) show that, in accordance with their phenology, bryophytes of this type have lower temperature optima for growth than the vascular plants with which they are associated in the field.

STRESS-TOLERANT COMPETITORS

A large number of plant species are typically associated with vegetation types which exhibit moderate productivity and experience very low intensities of disturbance. As we might expect, the characteristics of these plants fall between those of the competitors and the stress-tolerators. They may be conveniently classified into herbaceous and woody plants.

Herbaceous plants

In derelict grassland, heath, and marshland in temperate regions there are many grasses, sedges, and rushes which fall into the category of the stress-tolerant competitors. In common with the 'competitors' (e.g. *Urtica dioica, Epilobium hirsutum*) these plants are robust perennials which have the capacity for lateral vegetative spread by means of rhizomes or expanding tussocks. However, points of difference from the competitive herbs are the lower maximum potential relative growth-rates (Grime and Hunt, 1975), the longer life-span of the leaves, and the shoot phenology which in many species shows an exact compromise between that of the competitor and the stress-tolerator, i.e. they are evergreens with a marked decline in shoot biomass during the winter and a pronounced peak in the summer (Figure 15). European grasses which exemplify this strategy occur in a variety of habitats; they include *Brachypodium pinnatum, Bromus erectus, Festuca arundinacea, Festuca rubra* (semi-derelict grassland), *Molinia caerulea** (marshes), and *Ammophila arenaria* and *Elymus arenarius* (sand-dunes). In marshland there are numerous examples among both sedges, e.g. *Eriophorum vaginatum, Cladium mariscus, Scirpus sylvaticus*, and *Carex acutiformis* (Plate 11), and rushes, e.g. *Juncus effusus, J. acutus*, and *J. subnodulosus*. Among dicotyledonous herbs a most familiar example is provided by the woodland plant *Mercurialis perennis* (Plate 12). This shade-tolerant species is strongly rhizomatous and

***M. caerulea* is exceptional among the stress-tolerant competitors in that it is deciduous. In other respects, however, the species appears to conform to this category.

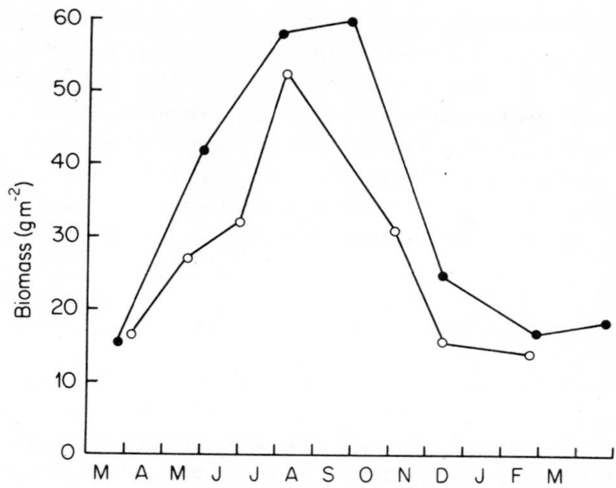

Figure 15 Seasonal change in the shoot biomass in the perennial grass, *Festuca rubra*, measured in derelict calcareous pastures at two sites in Northern (o–o) and Southern (•–•) England (Williamson, 1976; Al-Mufti *et al*., 1977).

forms extensive stands with a leaf canopy of moderate density in which individual leaves may remain alive from March to November. Another woodland species of rather similar ecology is the monocotyledon *Convallaria majalis*.

Stress-tolerant competitors are strongly represented among the grasses and sedges exploiting the semi-arid conditions of the North American prairies and the Russian Steppes, where typical species include *Stipa viridula, S. spartea, Sporobolus heterolepis*, and *Carex pennsylvanica* (Weaver and Albertson, 1956). Another species which appears to provide a particularly good example is *Carex lacustris*, a North American wetland species. Intensive studies of this plant (Bernard and MacDonald, 1974; Bernard and Solsky, 1977) have revealed a shoot phenology typical of the stress-tolerant competitor and a maximum life-span for the leaves of approximately one year.

Woody plants

As explained on page 149, there is reason to suspect that the majority of the trees and shrubs which occur in unproductive habitats or in the later stages of vegetation succession in more fertile terrain fall into the category of the stress-tolerant competitor. Because of the shortage of comparative data on the life histories and physiology of trees and shrubs there is insufficient evidence with which to examine this hypothesis. It is apparent, however, that many woody species exhibit, in particular attributes, degrees of specialization which lie between the two extremes recognized in Table 6 as characteristic of the competitor and the stress-tolerator. Examples here include the intermediate longevity of many deciduous forest trees in genera such as *Quercus, Carya, Castanea*, and *Fagus* and their fluctuating but fairly continuous seed output (Harper and White, 1974). A feature of certain deciduous shrubs, e.g. *Euonymus europaeus, Vaccinium myrtillus*, which may indicate their intermediate status is their capacity to maintain some photosynthetic activity throughout the year by the possession of evergreen stems (Perry, 1971). Until much more information is available, however, it is probably unwise to attempt to classify precisely with respect to strategy even the most common shrubs, woody climbing plants, and trees.

'C–S–R' STRATEGISTS

In certain environments it is possible to identify vegetation types in which plants conforming to very different strategies are growing together. In Chapter 6 the circumstances which allow this to occur will be analysed; here it will suffice to observe that co-existence of widely-different strategies is associated with habitats in which the respective effects of competition, stress, and disturbance are confined to particular times in the year and/or show local spatial variation in intensity within the habitat. However, not all of the habitats which experience major impacts of competition, stress, and disturbance are subject to pronounced seasonal or spatial variation in the equilibrium between the three phenomena.* In many unfertilized pastures in temperate regions, for example, mineral nutrient-stress and moderate intensities of defoliation by grazing animals are more or less constant features of the habitat. In these circumstances the bulk of the vegetation is usually composed of species with characteristics intermediate between those of the competitor, the stress-tolerator, and the ruderal. Table 9 contains a list of examples of 'C–S–R strategists' of common occurrence in unproductive pastures and grazed marshes of the British Isles. The majority of the grasses are perennials of rather small stature and moderate maximum potential relative growth-rate (e.g. *Festuca ovina* (Plate 28), *Nardus stricta, Helictotrichon* spp.). The sedges include a large number of small, shortly-rhizomatous species (e.g. *Scripus setaceus, Luzula campestris, Carex panicea*). A variety of growth-forms and phenologies are apparent in the dicotyledons, but the majority possess rosettes of leaves (e.g. *Succisa pratensis, Lotus corniculatus, Poterium sanguisorba*). As more information becomes available it seems inevitable that a range of types will be distinguished within the C–S–R strategists. As an illustration of the variety of substrategies, three types of common occurrence in unproductive grasslands of the British Isles will now be considered.

Small tussock grasses

Where pastures occur on shallow or sandy soils and are subject to mineral nutrient stress and possibly also to desiccation, a high proportion of the stand-

*It is worth noting, however, that even where the point of equilibrium remains more constant there may be considerable seasonal and spatial variation in types of stress and in agents of disturbance. This has important implications with regard to the variety of species maintained in the vegetation (see Chapter 6).

Table 9. Some C–S–R strategists of the British Flora classified into six sub-groups according to taxonomic and morphological criteria.

(a) Small tussock grasses

Festuca ovina	*Anthoxanthum odoratum*
Koeleria cristata	*Sesleria albicans*
Helictotrichon pratense	*Briza media*

(b) Small deep-rooted forbs with rosettes

Poterium sanguisorba	*Centaurea nigra*
Pimpinella saxifraga	*Scabiosa columbaria*
Leontodon hispidus	*Silene nutans*

(c) Small stoloniferous species

Fragaria vesca	*Hieracium pilosella*
Potentilla sterilis	*Veronica chamaedrys*
Ajuga reptans	*Glechoma hederacea*

(d) Forb with short rhizome or stock

Betonica officinalis	*Potentilla erecta*
Prunella vulgaris	*Chrysanthemum leucanthemum*
Succisa pratensis	*Serratula tinctoria*

(e) Legumes

Lotus corniculatus	*Trifolium medium*
Anthyllis vulneraria	*Astragalus danicus*
Lathyrus montanus	*Hippocrepis comosa*

(f) Small sedges and rushes

Carex caryophyllea	*Juncus bulbosus*
Scirpus setaceus	*Luzula campestris*
Juncus squarrosus	*Carex nigra*

ing crop is composed of tussock grasses with narrow, folded, or rolled leaves. Certain of these species, e.g. *Anthoxanthum odoratum, Festuca ovina*, and *Koeleria cristata*, retain green shoots throughout the year but also produce a well-defined peak in shoot weight which coincides with flowering in the late spring (Figure 16). Grasses of this type tend to be rather shallow in root penetration (Figure 17) and it is not uncommon for the tussocks to suffer partial or total mortality among the tillers during the summer.

Small tussock grasses are a ubiquitous feature of unproductive pastures. In the North American prairies, for example, the ecological niche which in Britain is occupied by *Festuca ovina* is filled by the closely-similar species *Bouteloua gracilis* which 'survives on ridges and dry uplands where few other grasses can grow' (Weaver and Albertson, 1956).

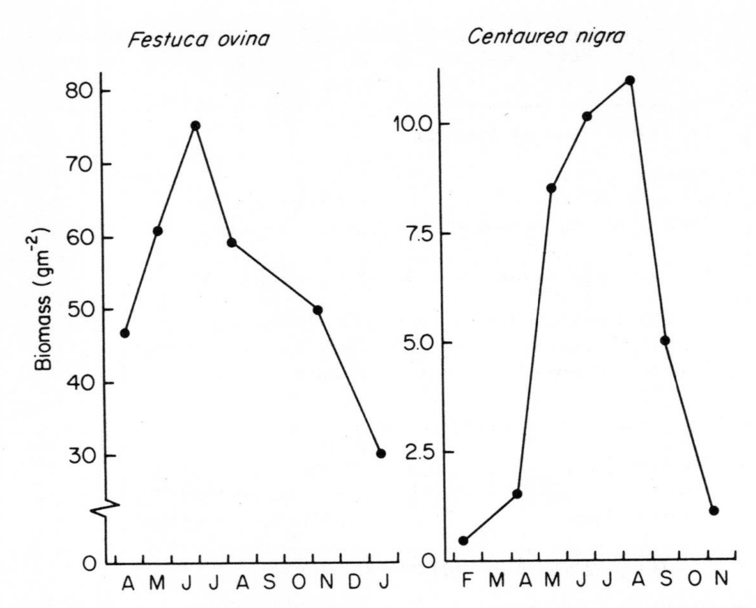

Figure 16 Seasonal change in shoot biomass in two contrasted C–S–R strategists of common occurrence in calcareous pastures in Europe. *Festuca ovina* is a small tussock grass; *Centaurea nigra* is a small rosette forb with a long tap-root (Al-Mufti *et al*., 1977). (Reproduced by permission of Blackwell Scientific Publications Ltd.)

Small, deep-rooted forbs

A second group which may be recognized is composed of forbs such as *Poterium sanguisorba*, *Centaurea nigra*, and *Leontodon hispidus*, the phenology of which resembles that of the competitors in that there is a well-defined summer peak in shoot production (Figures 16 and 55). However, in certain other respects these plants differ considerably from the competitors. In each, the shoot biomass is relatively small and is composed of one or more rosettes of leaves, a morphology which severely restricts both the height of the leaf canopy and the capacity of the shoot for lateral spread above and below ground. An additional characteristic of these plants is the presence of a long tap-root system. As Walter (1973) has pointed out, such roots usually penetrate deep fissures and allow the species to exploit reserves of moisture which are inaccessible to grasses and other shallow-rooted plants, many of which often tend to become severely desiccated during the summer.

An additional feature of many of these plants is the capacity for regeneration of the shoot from buds situated near the top of the tap-root. In relatively palatable species such as *Poterium sanguisorba*, *Leontodon hispidus*, and *Hypochoeris radicata*, this characteristic appears to play an important role in allowing the persistence of these plants in pastures subject to close grazing by sheep, rabbits, and invertebrates.

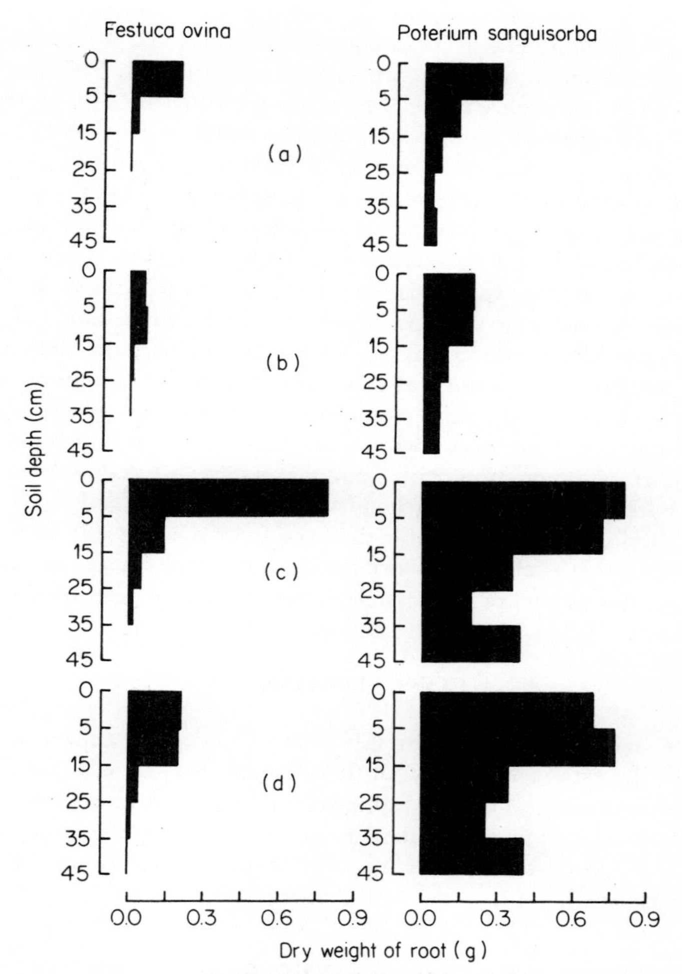

Figure 17 Root profiles of seedlings of two contrasted types of C–S–R strategists after six weeks' growth on an unfertilized soil removed from a site colonized by the two plants ((a) and (b)), and on the same soil supplemented by Hewitt nutrient solution ((c) and (d)). In treatments (b) and (d) the 0–5 cm zone of the soil profile was allowed to dry out during the last three weeks of the experiment (Sydes, unpublished).

Small, creeping or stoloniferous forbs

A recurrent morphology in certain pastures containing C–S–R strategists is that of the creeping plant.* In some species (e.g. *Veronica chamaedrys*, *Teucrium*

* Creeping growth forms are not, of course, peculiar to C–S–R strategists. When allied to other growth characteristics (see pages 11 and 60) rhizomes and stolons may assume a quite different strategic significance.

scorodonia) the creeping shoots bear leaves and occasional adventitious roots along their entire length, whilst in others (e.g. *Fragaria vesca, Hieracium pilosella, Rubus saxatilis*) the shoots are in the form of stolons which are capable of producing daughter rosettes, which although often situated at considerable distance from the parent plant remain connected at least during the first year of their existence. It is tempting to interpret these growth-forms mainly in relation to the process of vegetative propagation. However, without denying this function, another possibility should be considered. A characteristic of the ecology of many, if not all, of these species is the association with habitats such as rock outcrops, screes, and quarry heaps, in which a high proportion of the ground surface is covered by stone. In this type of habitat opportunities for rooting are extremely localized and dense leaf canopies tend to develop above the areas where soil is accessible. In such circumstances it is apparent that a selective advantage may accrue to species which can absorb water and mineral nutrients from one part of the environmental mosaic, and photosynthesize in another, i.e. plants with growth forms which cause leaves to be subtended over areas of bare rock situated at a considerable distance from the roots. It seems reasonable to suggest that such growth forms may allow exploitation of gaps in herbaceous canopies which are inaccessible to the majority of neighbouring herbs.*

CONCLUSIONS

From the evidence reviewed in this chapter it would appear that plant strategies in the established phase of the life-cycle may be classified by reference to a defined range of equilibria between stress, disturbance, and competition. This classification introduces more subtlety in the recognition of strategies and allows some observations to be made concerning the strategic range of particular life-forms and taxonomic groups.

The relationship between strategy and life-form

In Figure 18 an attempt has been made to describe the approximate strategic range of selected life-forms. The widest range of strategies is that attributed to perennial herbs and ferns. Annual herbs are predominantly ruderal whilst biennial herbs become prominent in the areas of the triangular model corresponding to the competitive-ruderals and the stress-tolerant ruderals. Trees and shrubs comprise competitors, stress-tolerant competitors, and stress-tolerators. Although lichens are confined to the stress-tolerant corner of the model, bryophytes are more wide-ranging with the centre of the distribution in the stress-tolerant ruderals.

* In passing it is interesting to observe that the same explanation can be applied on a larger scale to the creeping growth-forms associated with the pools and bare mud in marshes (*Ranunculus repens, Agrostis stolonifera*), woodland hollows filled with persistent tree litter (*Rubus fruticosus, Galeobdolon luteum*), and walls, cliffs, and tree trunks (*Hedera helix, Clematis vitalba*).

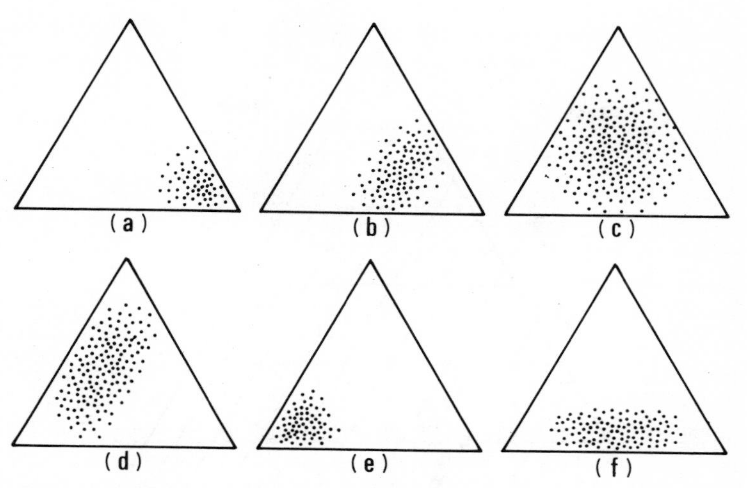

Figure 18 Diagrams describing the range of strategies encompassed by (a) annual herbs, (b) biennial herbs, (c) perennial herbs and ferns, (d) trees and shrubs, (e) lichens, and (f) bryophytes. For the distribution of strategies within the triangle, see Figure 12. (Reproduced from *American Naturalist*, **111**, by permission of the University of Chicago Press. ©1977. The University of Chicago Press.)

Triangular ordination

If the triangular model is an accurate general summary of the range of contingencies to which the established phase may be adapted then it should provide a basis upon which to classify both plants and vegetation types. In 1974 an attempt was made to explore this possibility using field survey data and growth analysis results based upon the herbaceous flora of the Sheffield region in Northern England. The method involved a triangular ordination in which species were classified with respect to two criteria. The first of these was the potential maximum rate of dry matter production (R_{max}) measured under a standardized productive environment, whilst the second was a morphology index* reflecting the maximum size attained by the plant under favourable conditions. This approach was based therefore upon the hypothesis that in herbaceous plants the primary strategies correspond to three permutations between R_{max} and morphology, i.e. rapidly-growing and large (competitors), rapidly-growing and small (ruderals), and slow-growing and small (stress-tolerators). Particularly in its use of provisional estimates of R_{max} based upon samples from single populations, this form of classification is extremely unsubtle. Despite these limitations, consistent patterns of distribution were obtained

*More recent studies suggest that, eventually, the morphology index (originally described inappropriately Grime, 1974 as a competitive index) may be replaced by a simpler index based upon the potential of the plant for lateral spread. On this basis, it would seem possible that a method of ordination may be devised which could be applied to woody species, herbs, and non-vascular plants.

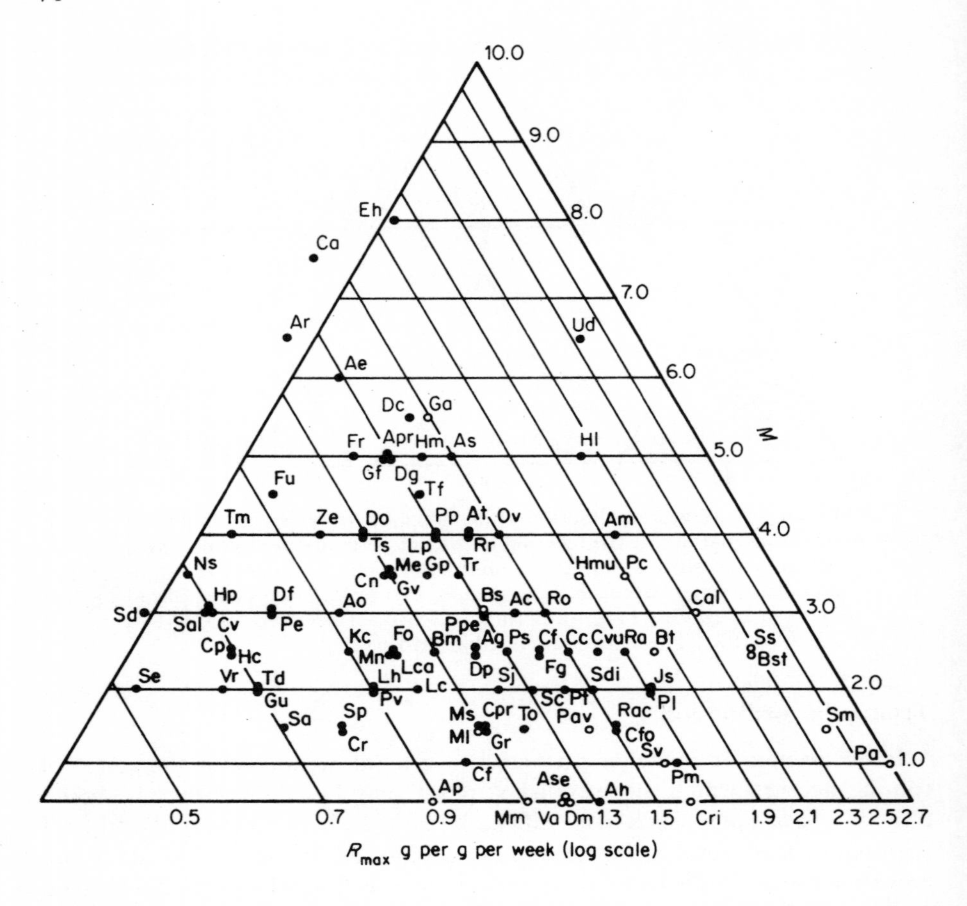

Figure 19 A triangular ordination of herbaceous species. ○, annuals; ●, perennials (including biennials). The morphology index (M) was calculated from the formula $M = (a + b + c)/2$ where a is the estimated maximum height of leaf canopy (1, <12 cm; 2, 12–25 cm; 3, 25–37 cm; 4, 37–50 cm; 5, 50–62 cm; 6, 62–75 cm; 7, 75–87 cm; 8, 87–100 cm; 9, 100–112 cm; 10, >112 cm); b is the lateral spread (0, small therophytes; 1, robust therophytes; 2, perennials with compact unbranched rhizome or forming small (<10 cm diameter) tussock; 3, perennials with rhizomatous system or tussock attaining diameter 10–25 cm; 4, perennials attaining diameter 26–100 cm; 5, perennials attaining diameter >100 cm); c is the estimated maximum accumulation of persistent litter (0, none; 1, thin discontinuous cover; 2, thin, continuous cover; 3, up to 1 cm depth; 4, up to 5 cm depth; 5, >5 cm depth) (Grime, 1974).

Key to species: Ac, *Agrostis canina*, ssp. *canina*; Ae, *Arrhenatherum elatius*; Ag, *Alopecurus geniculatus*; Ah, *Arabis hirsuta*; Am, *Achillea millefolium*; Ao, *Anthoxanthum ordoratum*; Ap, *Aira praecox*; Apr, *Alopecurus pratensis*; Ar, *Agropyron repens*; As, *Agrostis stolonifera*; Ase, *Arenaria serpyllifolia*; At, *Agrostis tenuis*, Bm, *Briza media*, Bs, *Brachypodium sylvaticum*; Bst, *Bromus sterilis*; Bt, *Bidens tripartita*; Ca, *Chamaenerion angustifolium*; Cal, *Chenopodium album*; Cc, *Cynosurus cristatus*; Cf, *Carex flacca*; Cfl, *Cardamine flexuosa*; Cfo, *Cerastium fontanum*, Cn, *Centaurea nigra*; Cp, *Carex panicea*; Cpr, *Cardamine pratensis*, Cr, *Campanula rotundifolia*; Cri, *Catapodium rigidum*; Cv, *Clinopodium vulgare*; Cvu, *Cirsium vulgare*; Dc, *Deschampsia cespitosa*; Df, *Deschampsia flexuosa*; Dg, *Dactylis glomerata*; Dm, *Draba muralis*; Do, *Dryas octopetala*;

(Figure 19) and, with few exceptions, species with strong ecological affinities were found to be located in close proximity to each other. The species which in Figure 19 are closest to the left-hand corner of the triangle include evergreens tolerant of desiccation (*Sedum acre, Thymus drucei, Helianthemum chamaecistus*), cold (*Dryas octopetala, Nardus stricta*), shade (*Viola riviniana, Sanicula europaea, Deschampsia flexuosa*), or frequent in mineral nutrient-deficient habitats (*Helictotrichon pratense, Sesleria albicans, Sieglingia decumbens*). As one might expect, annual plants of productive severely disturbed habitats such as arable land (*Poa annua, Stellaria media*) are located in the 'ruderal' corner and perennial herbs of fertile, derelict environments such as *Urtica dioica, Epilobium hirsutum, Chamaenerion angustifolium* occur towards the apex of the triangle.

Many of the species included in Figure 19 have been cited in earlier sections of this chapter as examples of particular secondary strategies, and it is interesting to observe that the majority of these plants also fall into the expected patterns of distribution. Typical examples here are the competitive-ruderals *Chenopodium album* and *Holcus lanatus*, and the stress-tolerant competitors *Festuca rubra* and *Bromus erectus*. The location of the small winter annuals *Aira praecox, Veronica arvensis, Arenaria serpyllifolia*, and *Draba muralis* coincides with that predicted for the stress-tolerant ruderals and is quite distinct from that of the C–S–R strategists such as *Festuca ovina* and *Koeleria cristata*, with which these annuals occur on limestone outcrops. This suggests that in such habitats the patches of bare soil occupied by the annuals constitute a distinct micro-environment (see page 167).

The second step in evaluating triangular ordination was to attempt to classify vegetation samples from a wide variety of habitats. The position of each square-metre sample in this ordination is derived from weighted means for R_{max}

Dp, *Digitalis purpurea*; Eh, *Epilobium hirsutum*; Fg, *Festuca gigantea*; Fo, *Festuca ovina*; Fr, *Festuca rubra*; Fu, *Filipendula ulmaria*; Ga, *Galium aparine*; Gf, *Glyceria fluitans*; Gp, *Galium palustre*; Gr, *Geranium robertianum*; Gu, *Geum urbanum*; Gv, *Galium verum*; Hc, *Helianthemum chamaecistus*; Hl, *Holcus lanatus*; Hm, *Holcus mollis*; Hmu, *Hordeum murinum*; Hp, *Helictotrichon pratense*; Js, *Juncus squarrosus*; Kc, *Koeleria cristata*; Lc, *Lotus corniculatus*; Lca, *Luzula campestris*; Lh, *Leontodon hispidus*; Lp, *Lolium perenne*; Me, *Milium effusum*; Ml, *Medicago lupulina*; Mm, *Matricaria matricarioides*; Mn, *Melica nutans*; Ms, *Myosotis sylvatica*; Ns, *Nardus stricta*; Ov, *Origanum vulgare*, Pa, *Poa annua*, Pav, *Polygonum aviculare*; Pc, *Polygonum convolvulus*; Pe, *Potentilla erecta*; Pl, *Plantago lanceolata*; Pm, *Plantago major*; Pp, *Poa pratensis*; Ppe, *Polygonum persicaria*; Ps, *Poterium sanguisorba*; Pt, *Poa trivialis*; Pv, *Prunella vulgaris*; Ra, *Rumex acetosa*; Rac, *Rumex acetosella*; Ro, *Rumex obtusifolius*; Rr, *Ranunculus repens*; Sa, *Sedum acre*; Sal, *Sesleria albicans*; Sc, *Scabiosa columbaria*; Sd, *Sieglingia decumbens*; Sdi, *Silene dioica*; Sj, *Senecio jacobaea*; Sm, *Stellaria media*; Sp, *Succisa pratensis*; Ss, *Senecio squalidus*; Sv, *Senecio vulgaris*; Td, *Thymus drucei*; Tf, *Tussilago farfara*; Tm, *Trifolium medium*; To, *Taraxacum officinalis*; Tr, *Trifolium repens*; Ts, *Teucrium scorodonia*; Ud, *Urtica dioica*; Va, *Veronica arvensis*; Vr, *Viola riviniana*; Ze, *Bromus erectus*.

Estimates of R_{max} are based on measurements during the period 2–5 weeks after germination in a standardized productive controlled environment conducted on seedlings from seeds collected from a single population in Northern England (Grime, 1974). (Reproduced by permission of Macmillan (Journals) Ltd.)

and for the morphology index based upon all the species present in the vegetation sample.

A selection of the ordinations is presented in Figure 20a–o. The various types of vegetation occupy positions in close agreement with those which may be predicted from Figure 12. The distributions in Figure 20 illustrate the tendency of samples from stressed habitats (a, b, c), disturbed environments (j, k), and semi-derelict but productive sites (g, h, i) to extend into respective corners of the triangle. Rather unproductive vegetation types experiencing a moderate intensity of orderly disturbance (d, e, f) tend to occupy compact areas in the centre of the diagram. By contrast, samples from spoiled land (l–o) show an attenuated distribution which seems to represent the course of vegetation succession in these new habitats.

The consistent patterns evident in the results of this rudimentary 'classification by strategy' encourage the view, originating from the studies of Raunkiaer (1934), and sustained and expanded by Ellenberg (1963) and Ellenberg and Mueller-Dombois, (1967a, b), that future methods of vegetation analysis and description will rely increasingly upon criteria which are functional (i.e. concerned with characteristics of life-history and physiology) rather than taxonomic.

Intraspecific variation with respect to strategy

There is now a large quantity of published evidence of genetic variation within plant species, and it is clear that this variation may occur both between populations and within them. In our present state of knowledge it is not possible to draw a general perspective with regard to either the plant characteristics, which are most commonly subject to intraspecific variation, or to the circumstances in which the phenomenon exercises a major effect on the ecological amplitude of the species. It is already apparent, however, that in specific instances (e. g. Böcher, 1949) genetic variation is sufficient to enlarge substantially the strategic and ecological range of the species. From the investigations of Law *et al.* (1977) and Law (1978), for example, it is evident that within the common grass, *Poa annua*, there are populations which differ considerably in life-history. In severely disturbed habitats *P. annua* occurs as ephemeral plants which are typical representatives of the ruderal strategy. In marked contrast (Plate 13), populations of the same species in productive pastures contain a high proportion of biennial or possibly even perennial plants which may be described more accurately as competitive-ruderals.

A second example of intraspecific variation with respect to strategy is available from several studies conducted on the perennial grass *Agrostis tenuis*. From investigations such as those of Jowett (1964) it is apparent that whilst pasture populations are usually composed of potentially fast-growing plants of moderately high competitive ability, the species is represented on infertile mine-waste by stress-tolerant individuals of smaller stature and slower potential growth-rate.

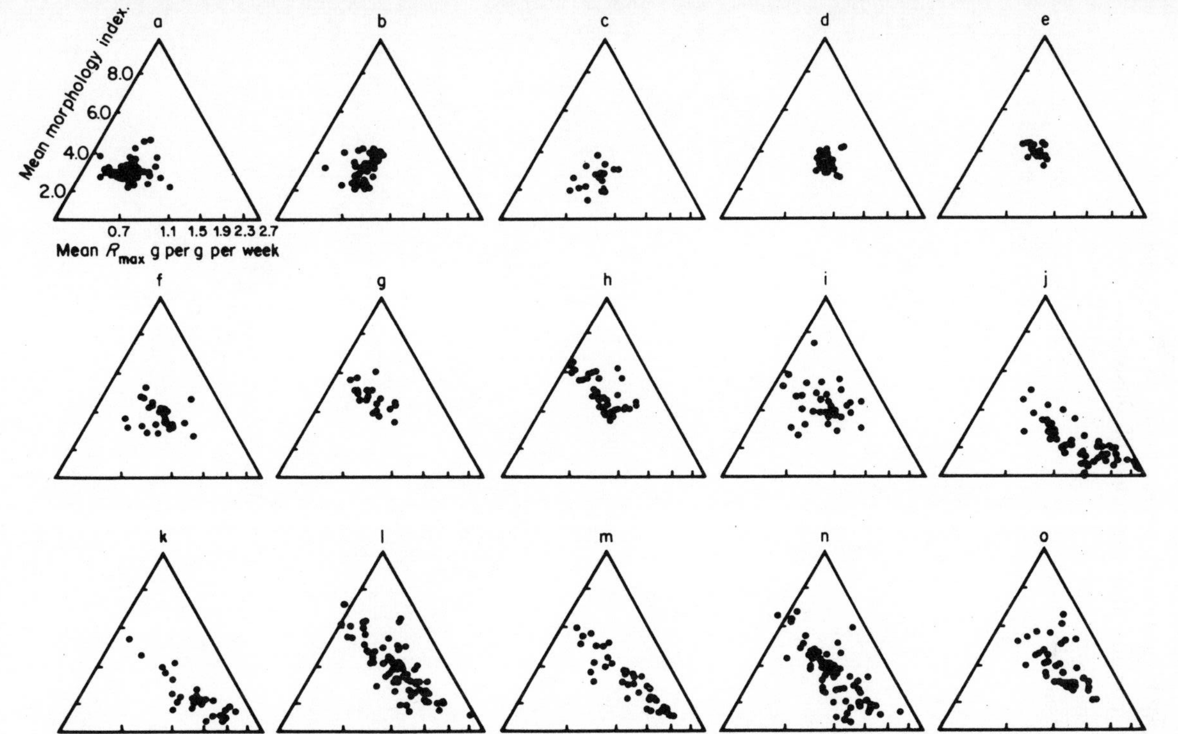

Figure 20 Triangular ordinations of m² samples of herbaceous vegetation from fifteen habitats. Axes are mean morphology index and mean R_{max} each derived as in Figure 19 and weighted according to the relative frequency of the species in the sample. (a) Unenclosed sheep pastures on acidic strata; (b) unenclosed sheep pastures on limestone; (c) limestone outcrops, (d) meadows; (e) road verges, mown frequently; (f) enclosed pastures; (g) road verges, mown infrequently; (h) hedge bottoms; (i) derelict banks of rivers, ponds, and ditches; (j) paths; (k) fallow arable; (l) heaps of mineral soil (such as building sites); (m) demolition sites (brick and mortar rubble); (n) cinders (tips and railway ballast); (o) manure heaps and sewage sludge (Grime, 1974). (Reproduced by permission of Macmillan (Journals) Ltd.)

Further evidence of the development of stress-tolerant populations in species more usually associated with mesic environments has been obtained from the studies of Böcher and Larsen (1958) and Böcher (1961) from which it is apparent that, in tundra conditions, species such as *Holcus lanatus* and *Dactylis glomerata* may have considerably extended life-histories.

Chapter 3

Regenerative strategies

INTRODUCTION

There is enormous variety in the mechanisms whereby plants regenerate, and there can be little doubt that this fact accounts for many of the differences in ecology observed between species. It follows therefore that in order to understand fully the ecology of a plant species or population it is necessary to characterize the strategies adopted in both the regenerative (immature) and the established (mature) phases of the life-cycle. Often the analysis is complicated even further by the fact that plants may possess more than one method of regeneration. In the concluding parts of this chapter the effect of particular strategies on the ecology of plant species will be considered. First, however, we must attempt to recognize the main types of regenerative strategies in plants and here a necessary preliminary is to examine the distinction between vegetative reproduction and regeneration by seed.

REGENERATION BY VEGETATIVE OFFSPRING AND BY SEED

In many plants, population expansion is capable of occurring, at least in the short term, without the involvement of sexual methods of regeneration. The majority of perennial herbs, ferns, and bryophytes together with some trees and shrubs are able to produce new individuals either by proliferation and subsequent fragmentation of the plant or through the formation of vegetative propagules such as bulbils, tubers, or gemmae. This type of regeneration is associated particularly with local consolidation of populations and has the effect of producing populations of genetically-uniform plants.

Consequences of exclusively vegetative regeneration often include, in addition to genetic uniformity, some restriction of the ecological range of the plant and a tendency for the populations to respond catastrophically to habitat changes induced by factors such as climate, predation, and vegetation management. It is not, perhaps, surprising therefore to find that plant populations rarely, if ever, depend exclusively upon vegetative reproduction; later in this chapter an attempt will be made to examine the complementary roles of vegetative and sexual reproduction in certain life-histories.

The importance of sexual reproduction derives only in part from the basis which the resulting genotypic variety provides for the responses of populations and species to natural selection. In flowering plants it is also characterized by the peculiar properties of seeds. Relative to the majority of vegetative propagules, seeds are numerous, independent, and stress-tolerant, and these characteristics confer, respectively, the potential for rapid multiplication, dispersal, and dormancy.

Seedlings of the majority of plants are small in size and many are incapable of survival in the conditions experienced by established plants of the same genotype. The dependence of many species upon local and unusually favourable sites for seedling establishment has been known for a considerable period of time and a comprehensive review of this phenomenon has been assembled (Grubb, 1977). It is evident that many plants possess mechanisms of seed production, dispersal, and dormancy which, together with the germination requirements, facilitate seedling establishment in circumstances which are quite distinct from those experienced during the later stages of the life-cycle and are local and/or intermittent in occurrence. Moreover, in long-lived or vegetatively-reproducing species, only low rates of regeneration from seed may be adequate for the maintenance of populations and in these circumstances populations may remain viable despite the fact that opportunities for seedling establishment occur relatively infrequently.

It is clear, therefore, that in terms of their capacity to influence the size, dispersion, survival, and genetic adaptability of plant populations, some broad distinctions can be drawn between the processes of vegetative reproduction and regeneration from seed. However, it is not possible to base a functional classification of regenerative strategies upon a simple dichotomy between vegetative reproduction and regeneration by seed. When the full range of plants is examined it is clear that a great deal of convergent evolution in structure and physiology has occurred between the two forms of regeneration, with the result that it is appropriate to include examples from both categories in the description of certain regenerative strategies.

TYPES OF REGENERATIVE STRATEGY

Information concerning mechanisms of regeneration by native plants is extremely fragmentary and it would be unwise to attempt a comprehensive and detailed account of this subject. However, certain basic forms of regeneration are now recognizable, and these may be provisionally classified into five types.

Vegetative expansion (*V*)

Under this heading it is convenient to assemble many of the regenerative mechanisms which involve the expansion and subsequent fragmentation of the vegetative plant through the formation of persistent rhizomes, stolons, or suckers. The most consistent feature of vegetative expansion is the low risk of mortality to the offspring. This is achieved through prolonged attachment to the parent plant and mobilization of resources from parent to offspring in quantities adequate to sustain the offspring during establishment.

Vegetative expansion is associated particularly with the maintenance and local consolidation of populations of perennial herbs and shrubs. Because of the lower risks of mortality this regenerative strategy is often successful in vegetation types in which establishment from seed is precluded by the presence of a dense cover of vegetation and litter. In temperate regions vegetative expansion is very common among tall herbs, e.g. *Epilobium hirsutum*, *Chamaenerion angustifolium*, and *Urtica dioica*, and also allows certain woody plants, e.g. *Populus tremula*, *Prunus spinosa*, *Hippophae rhamnoides*, and *Symphoricarpos rivularis*, to form extensive thickets in relatively undisturbed environments.

Regeneration by vegetative expansion, although less conspicuous, is also of frequent occurrence in unproductive habitats. As Billings and Mooney (1968) and Callaghan (1976) have recorded, it is particularly common in arctic and alpine environments and northern heathlands, where species such as *Gaultheria shallon*, *Salix herbacea*, *Dryas octopetala*, *Vaccinium myrtillus*, *V. vitis-idaea*, and *Carex bigelowii* may develop clonal patches of a considerable size. Vegetative sprouting is also a common form of regeneration in circumstances where shrubs are subject to intermittent damage by fire (Hanes, 1971; Hansen, 1976). In the coastal and desert chaparral of California, for example, vegetative expansion is evident in dominant species such as *Adenostoria fasciculatum* and *Ceanothus leucodermis*. Another vegetation type of low productivity in which vegetative expansion plays a most important role is tropical rain-forest. From a number of sources there is evidence (see page 181) that where local disturbance causes relatively small openings in the forest canopy, vegetative sprouts from neighbouring established trees usually provide the major source of invading plants.

In many of the herbs and woody species which exhibit vegetative expansion the breakdown of the connections between parent and offspring is delayed for many years, and it is often exceedingly difficult to draw a distinction between these plants and the large number of trees and shrubs in which vegetative reproduction does not normally occur.

In vegetative expansion, a low risk of mortality to the offspring is achieved through attachment to the parent and a necessary condition for success is the survival of the parent during the period of establishment. In rhizomatous grasses and forbs in productive habitats establishment may be completed within one growing season, whereas in stressed environments, and particularly in woody plants, the process may extend over many years. It is hardly surprising, therefore, to find that vegetative expansion is most apparent in relatively undisturbed habitats. In contrast, the remaining strategies to be described are all adapted to exploit disturbance and differ from each other according to the nature and frequency of this disturbance and the environmental context in which it occurs.

Seasonal regeneration in vegetation gaps (*S*)

In a wide range of habitats, herbaceous vegetation is subjected to seasonally-predictable damage by phenomena such as temporary drought, flooding, trampling, and grazing. Under these conditions, the most common regenerative strategy is that in which areas of bare ground or sparse vegetation cover are created every year and are recolonized annually during a particularly favourable season. This form of regeneration reaches its highest frequency in the temperate zone where two main types can be distinguished with respect to the season in which recolonization occurs. In the first, establishment takes place in the autumn whereas in the second it is delayed until spring.

Autumn regeneration

In regions in which rainfall is mainly restricted to a cool season, herbaceous vegetation is subjected to highly predictable damage by drought; this effect is most pronounced in grasses, the majority of which are shallow-rooted. In many grasslands there is additional seasonal disturbance resulting from trampling and defoliation by wild or domestic animals.

In habitats of this type detached viable seeds of the majority of the grasses are present in the habitat only during the dry season, and sampling of the surface soil at intervals throughout the year (Figure 21) reveals a strong annual fluctuation in seed content. The numbers of germinable seeds reach a minimum during the wet season, at the beginning of which synchronous germination of the entire seed population results in the appearance of large numbers of seedlings of grasses in the areas of bare ground developed during the preceding dry season.

Grasses of this type possess relatively large seeds which are characterized by lack of innate dormancy and by the ability to germinate in light and in darkness over a wide range of temperatures (Table 10). All these characteristics are consistent with the conclusion that in these species moisture supply is the overriding determinant of the timing of germination. It is interesting to note that many of the commonest grasses of pastures and meadows in Europe (e.g. *Arrhenatherum elatius*, *Bromus erectus*, *Bromus sterilis*, *Cynosurus cristatus*, *Dactylis glomerata, Festuca pratensis, Festuca rubra, Hordeum murinum, Lolium multiflorum, Lolium perenne*) conform to this type. The failure of these species to develop persistent reserves of buried seeds in the soil is well-documented (Brenchley and Warington, 1930; Chippendale and Milton, 1934; Milton, 1939; Champness and Morris, 1948; Thompson and Grime, 1978) and is of profound importance to pasture management in Europe since it appears to play a crucial role in the process whereby sown species such as *Lolium perenne* and *L. multiflorum* tend to be replaced by native grasses (see page 96).

Although the most familiar type of autumn regeneration in seasonally droughted habitats involves the seeds of grasses it is interesting to note that there are similar mechanisms which depend upon the production of vegetative propagules. An example of this phenomenon is provided by *Allium vineale*, a

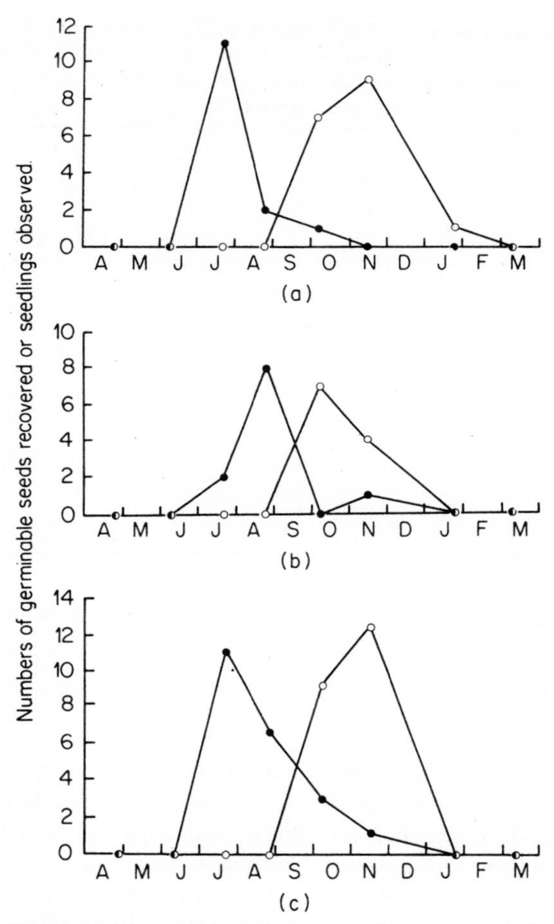

Figure 21 Seasonal variation in the numbers of detached viable seeds (●–●) and newly-germinated seedlings (○–○) of three tussock grasses in dry calcareous grassland in Northern England. (a) *Briza media* (b) *Helictotrichon pratense* (c) *Koeleria cristata*. Seed number refers to seeds germinating from a standard volume of surface (0–3 cm) soil removed to a controlled environment. Seedling number refers to freshly germinated seedlings (i.e. seedlings were removed after each count) in field plots of total area 0.2 m^2 (Thompson, 1977).

species in which regeneration is effected by bulbils which remain dormant until the onset of moist conditions in the autumn.

In addition there are less specialized mechanisms of vegetative regeneration whereby established plants which have been damaged by drought and/or predation during the summer may be rehabilitated in the autumn and winter. Examples here include the grass *Poa trivialis* and various bryophytes (e.g.

Table 10. Estimations of mean seed weight and germination response to constant temperatures in the light in British populations of ten grasses which are predominantly autumn germinators and do not develop buried seed banks (Grime & Rodman, unpublished).

Species	Mean seed weight (mg)	Temperature range for 50% germination	
		Lower limit	Upper limit
Arrhenatherum elatius	2.4	<5	29
Bromus erectus	4.2	10	34
Bromus sterilis	8.4	<5	37
Cynosurus cristatus	0.7	<5	33
Dactylis glomerata	0.5	<5	31
Festuca ovina	0.4	6	33
Festuca rubra	0.8	6	35
Hordeum murinum	6.6	<5	33
Lolium perenne	1.8	<5	34
Phleum pratense	0.4	11	36

Brachythecium rutabulum, *Eurhynchium praelongum*) in which new individuals originate from surviving meristems.

Spring regeneration

In northern or more continental regions within the temperate zone, plant growth in the late autumn and winter is often severely restricted by low temperatures and frost damage, and for this reason gaps in the vegetation which originated during the summer are likely to remain open until the following spring. Moreover, in many habitats, additional patches of bare soil arise during the winter and early spring as a result of frost-heaving, solifluction, and water erosion following snow-thaw. It is not surprising, therefore, to find that in plants of more northern latitudes there is a tendency for regeneration to be delayed until the spring.

In north-temperate and cold climates many trees, shrubs, and herbs are incapable of germination until the imbibed seed has experienced temperatures in the approximate range of 2–10°C. The length of the required chilling period varies from several days (e.g. *Impatiens glandulifera*) to months (e.g. *Impatiens parviflora*).

Table 11 contains a list of the species which were found to have a chilling-requirement when a screening of germination characteristics was carried out on the common plants present in a local flora in Northern England. In a small number of plants (e.g. *Polygonum aviculare*, *Galium palustre*) the chilling requirement appears to be associated with characteristics which cause many of the seeds to become incorporated into a buried seed bank (see page 92). However, for the majority of these species there is evidence from laboratory and/or field

Table 11. Some flowering plants for which there is evidence in British populations of a chilling requirement for germination (Grime, Curtis & Rodman, unpublished). The values in brackets refer to the mean seed weight (mg) of the populations investigated.

(a) Woodland plants

WOODY SPECIES		HERBACEOUS SPECIES	
Acer pseudoplatanus	(33.47)	*Allium ursinum*	(4.78)
Clematis vitalba	(1.27)	*Bromus ramosus*	(7.37)
Corylus avellana	(63.68)	*Campanula latifolia*	(0.06)
Frangula alnus	(17.05)	*Conopodium majus*	(2.26)
Fraxinus excelsior	(46.19)	*Endymion non-scriptus*	(6.17)
Ligustrum vulgare	(14.19)	*Festuca gigantea*	(3.12)
Lonicera periclymenum	(5.21)	*Galeobdolon luteum*	(1.98)
Prunus spinosa	(118.13)	*Impatiens parviflora*	(6.91)
*Quercus petraea**	(744.70)	*Mercurialis perennis*	(2.17)
Rosa pimpinellifolia	(15.84)	*Oxalis acetosella*	(0.99)
Sambucus nigra	(1.90)	*Sanicula europaea*	(2.97)
Sorbus aucuparia	(2.58)	*Viola riviniana*	(1.01)

(b) Plants of grassland, heathland and other types of herbaceous vegetation in unshaded habitats

Aegopodium podagraria	(2.73)	*Myrrhis odorata*	(35.01)
Agrimonia eupatoria	(23.78)	*Odontites verna*	(0.15)
Alchemilla vestita	(0.46)	*Pimpinella major*	(2.12)
Anthriscus sylvestris	(5.18)	*Pimpinella saxifraga*	(1.19)
Chaerophyllum temulentum	(1.44)	*Rhinanthus minor agg.*	(2.84)
Empetrum nigrum	(0.75)	*Smyrnium olusatrum*	(20.46)
Euphrasia officinalis	(0.13)	*Torilis japonica*	(1.98)
Heracleum sphondylium	(5.52)	*Viola hirta*	(2.79)

(c) Marsh plants

Angelica sylvestris	(1.15)	*Pinguicula vulgaris*	(0.02)
Caltha palustris	(0.99)	*Saponaria officinalis*	(1.46)
Eleocharis palustris	(0.96)	*Silaum silaus*	(1.09)
Galium palustre	(0.91)	*Solanum dulcamara*	(1.49)
Impatiens glandulifera	(7.32)		

(d) Arable weeds

Aethusa cynapium	(1.07)	*Papaver rhoeas*	(0.09)
Anagallis arvensis	(0.40)	*Polygonum aviculare*	(1.45)
Atriplex hastata	(0.86)	*Polygonum convolvulus*	(1.28)
Atriplex patula	(1.33)	*Polygonum hydropiper*	(1.24)
Chenopodium album	(0.77)	*Viola arvensis*	(0.40)
Papaver dubium	(0.12)		

*Chilling requirement specific to plumule extension.

investigations that germination of the entire populations of seeds occurs soon after the requirement for chilling has been satisfied. Seeds of these plants differ in several respects from those characteristic of buried seed banks (Table 17). The majority are quite large, none is inhibited from germination by darkness, and once the chilling requirement is fulfilled a high proportion is capable of germinating rapidly at low (<10°C) temperatures. The main consequence of this physiology is that the seeds are prevented from germinating during the summer and autumn and exhibit synchronous germination at a time in the winter or early spring determined by the length of the chilling requirement and the temperatures prevailing in the habitat.

Trees, shrubs, and herbs of deciduous woodlands, scrub, and hedgerows form a high proportion of the species listed in Table 11, and it seems likely that the main advantage of the chilling requirement in such plants is related to the fact that germination is caused to occur during the unshaded phase in the early spring at which time conditions are most favourable for seedling establishment. This phenomenon is particularly obvious in British woodlands where observations prior to the expansion of the tree canopy in the early spring reveal large, even-aged populations of germinating seedlings of woody species (e.g. *Acer pseudoplatanus (Plate 14)*, *Fraxinus excelsior*) and herbs (e.g. *Endymion non-scriptus* (Plate 15), *Galeobdolon luteum*). Evidence of the importance of early germination in woodland plants is available from studies such as those of Wardle (1959) and Gardner (1977) which have included reports of exceedingly high rates of seedling mortality during the summer maximum in shading by the canopy of trees and herbs.

A similar principle may be applied to explain the presence of a chilling requirement in many of the commonest forbs of productive herbaceous vegetation types such as meadows, roadsides, and river banks. In such habitats the density of living plant material and litter is at a minimum in the early spring and germination at this time allows seedling growth to precede that of the established perennials. In Europe, numerous examples of this phenomenon occur in annuals (e.g. *Impatiens glandulifera*, *Galeopsis tetrahit*, *Galium aparine*), biennials (e.g. *Anthriscus sylvestris*, *Heracleum sphondylium*, *Torilis japonica*), and perennials (e.g. *Myrrhis odorata*, *Pimpinella major*, *Silaum silaus*). In all these species the early spring germination is allied to a comparatively large seed and it seems likely that both contribute to the impetus to seedling growth which enables these plants to regenerate in close proximity to established perennials.

All of the examples considered so far have involved regeneration by seed and most of the species concerned have been annuals or biennials. However, we may also include in this category a large number of perennial herbs in which spring regeneration is a vegetative process. Although in certain of these plants propagation is by means of specialized structures such as the tubers of *Ranunculus ficaria*, in the majority of species new individuals originate each spring from fragments of rhizomes or stolons which were produced during the previous growing season and have remained dormant during the winter. The latter form of regeneration occurs in both grasses (e.g. *Agrostis stolonifera*, *Glycera fluitans*, *Poa pratensis*) and forbs (e.g. *Tussilago farfara*, *Cirsium arvense*, *Trifolium repens*) and differs from vegetative expansion in that (1) establishment of the offspring usually coincides with (or is preceded by) the death or senescence of the parent and (2) there is a higher rate of mortality among the offspring.

General characteristics of seasonal regeneration

A feature common to all the plants which exhibit seasonal regeneration is the capacity to invade vegetation gaps at a specific period in the year, and so far in this account attention has been concentrated on the mechanisms which control the timing of regeneration. However, an essential characteristic of vegetation gaps is their localization in space and it is interesting to observe that most of the plants which belong to this category possess mechanisms of regeneration which appear to increase the chances of 'gap-detection'.

In species in which seasonal regeneration is effected by seed, the flowering shoot is usually a large structure which is either compound and spreading (e.g. the majority of *Umbelliferae*) or tall and flexuous (e.g. *Bromus sterilis*, *Bromus erectus*). The most likely adaptive value of such inflorescences is that they ensure that, despite their large size, the seeds of these plants become dispersed within the habitat to an extent which ensures that a proportion fall onto bare soil or into areas where the density of established vegetation and litter is insufficient to prevent establishment. In certain herbaceous species such as *Impatiens glandulifera* and *Geranium robertianum* the same effect is achieved by means of explosive dehiscence of the seeds.

An essentially similar mechanism of gap-exploitation can be recognized in many of the species (e.g. *Tussilago farfara*, *Trifolium repens*, and *Ranunculus repens*) in which seasonal regeneration is primarily a vegetative process. In these plants, however, the dispersal of propagules is achieved by the proliferation and fragmentation of rhizomes and stolons.

It seems reasonable to conclude that the strategy of seasonal regeneration represents a rather unsophisticated mechanism of gap exploitation in which, after propagules have been dispersed within the habitat and have germinated simultaneously (or recommenced growth, in the case of fragments of rhizomes or stolons), survival is limited to those individuals which, by chance, occur in gaps. It is evident that this method of regeneration depends upon the creation each year of a high density of suitable gaps. As we shall see from later sections of this chapter, in vegetation in which gaps occur more rarely, seasonal regeneration gives way to more discriminating types of regenerative strategy.

Regeneration involving a persistent seed bank (B_S)

When flowering plants are compared with respect to the fate of their seeds, two contrasting groups may be recognized. In one, most if not all of the seeds germinate soon after release whilst in the other group many become incorporated into a bank of dormant seeds which is detectable in the habitat at all times during the year and may represent an accumulation of many years. These two groups are, of course, extremes and between them there are species and populations in which the seed bank, although present throughout the year, shows pronounced seasonal variation in size. Nevertheless, as indicated in Figure 22, it is convenient to draw an arbitrary distinction between 'transient' and 'persistent' seed banks. A transient seed bank may be defined as one in which none of the seed output remains in the habitat in a viable condition for more than one year. In the persistent seed bank, some of the component seeds are at least one year old.

Another distinction which may be drawn between transient and persistent seed banks relates to the location of the seeds within the habitat. Those workers who have split up the soil profile into horizons of different depths (Chippendale and Milton, 1934) or who have excluded surface soil (Wesson and Wareing, 1969a) before estimating the germinable seed content have demonstrated conclusively that the majority of persistent banks consist of buried seeds.

In order to examine the role of persistent seed banks in regeneration it will be helpful to identify the range of habitats with which they are associated.

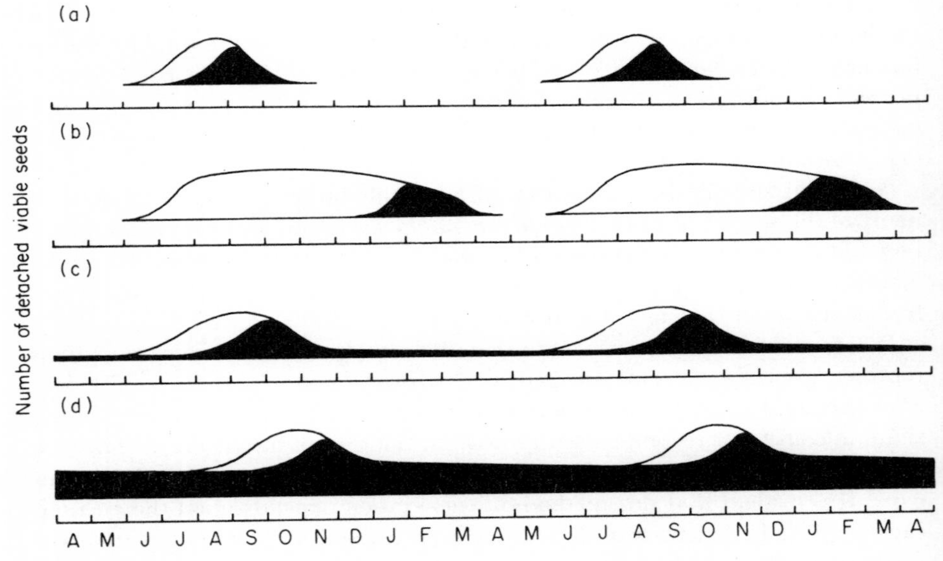

Figure 22 Scheme describing four types of seed banks of common occurrence in temperate regions. ■, seeds capable of germinating immediately after removal to suitable laboratory conditions. □, seeds viable but not capable of immediate germination. (a) Annual and perennial grasses of dry or disturbed habitats (e.g. *Hordeum murinum*, *Lolium perenne*, *Catapodium rigidum*); (b) annual and perennial herbs, shrubs, and trees colonizing vegetation gaps in early spring (e.g. *Impatiens glandulifera, Anthriscus sylvestris, Acer pseudoplatanus*); (c) winter annuals mainly germinating in the autumn but maintaining a small seed bank (e.g. *Arenaria serpyllifolia*, *Saxifraga tridactylites*, *Erophila verna*); (d) annual and perennial herbs and shrubs with large, persistent seed banks (e.g. *Stellaria media*, *Origanum vulgare*, *Calluna vulgaris*).

The distribution of persistent seed banks in relation to habitat

Where an attempt is to be made to characterize the seed banks of the various species present in a habitat it is necessary to sample the soil and litter at intervals throughout the year and to carry out germination procedures, or laboratory tests, to estimate seasonal variation in the numbers of viable seeds. The extremely laborious techniques which are required for accurate estimations of the numbers of seeds in soil samples have been described by Kropac (1966), Major and Pyott (1966), and Roberts (1970) and these are impracticable for large-scale studies, especially those involving small-seeded species. However, where the purpose is merely to detect the presence of a persistent seed bank in the soil it is not necessary to adopt such exhaustive methods. Data such as those illustrated in Figure 23, which were obtained by a simplified procedure of seasonal sampling and germination tests, provide ample confirmation of the presence of viable seeds throughout the year.

The best-documented examples of species which accumulate large reservoirs of seeds in the soil are the ruderals of arable fields (Brenchley, 1918; Brenchley

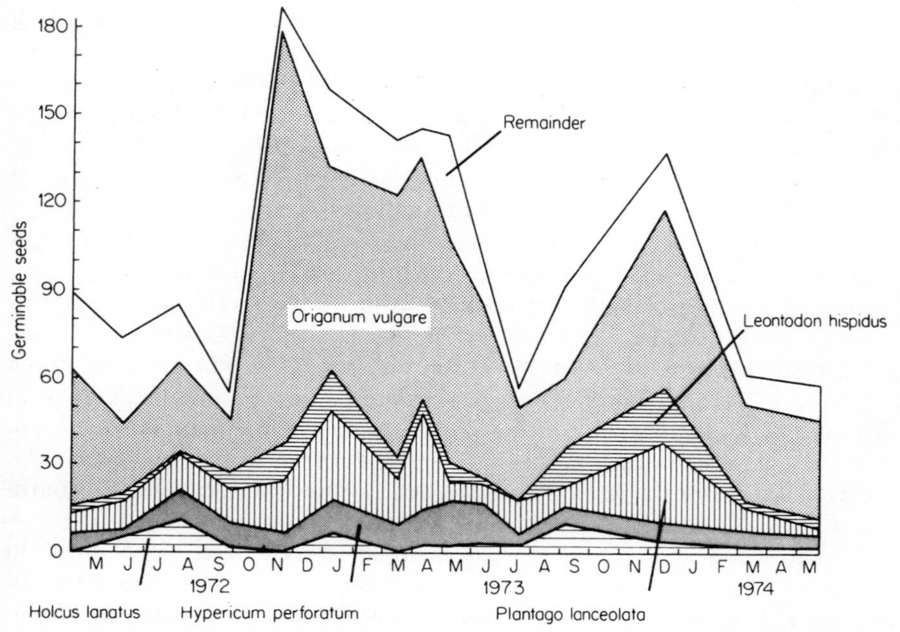

Figure 23 Estimations of seasonal variation in the numbers of detached germinable seeds present in a community of tall herbs in Northern England. The vegetation at the site was ungrazed but subject to occasional fires, none of which occurred during the period of the investigation. Estimations of the 'germinable seeds' are based upon the number of seedlings recorded during a standardized laboratory test on samples of soil (total volume 900 cm^3, surface area 450 cm^2) prepared by thorough mixing of subsamples removed from the soil surface (0–3 cm) at random positions within the site (Spray and Grime, in Grime, 1978).

and Warington, 1930; Barton, 1961; Roberts and Stokes, 1966; Roberts, 1970). For these plants there is abundant evidence of the capacity of buried seeds to survive for long periods and to germinate in large numbers when the habitat is disturbed. The advantage of the persistent seed bank is obvious in ruderal species of the intermittently-open habitats of farmland, since long-term survival of their populations frequently depends upon the ability to remain in a dormant condition through periods in which the habitat is occupied by a closed cover of perennial species. However, persistent seed banks are not confined to the ruderals of arable land; they occur in a wide range of ecological groups.

Large seed banks are particularly characteristic of shrubs (e.g. *Calluna vulgaris*) and perennial herbs (e.g. *Hypericum perforatum*, *Origanum vulgare*) occurring in habitats subjected to intermittent damage by fire. In a very different context, enormous banks of buried seeds are produced by some marshland species (e.g. *Juncus effusus*, *Juncus articulatus*, *Epilobium hirsutum*). Persistent seed banks are also characteristic of certain pasture plants (e.g. *Agrostis tenuis*, *Anthoxanthum odoratum*, *Holcus lanatus*, *Plantago lanceolata*) and they are especially common in species colonizing seasonally water-logged grasslands (e.g. *Deschampsia cespitosa*, *Poa trivialis*, *Ranunculus repens*, *Potentilla erecta*). Although some of the largest and more persistent seed banks are to be found in wetland habitats (Darwin, 1859; Milton, 1939; van der Valk and Davis, 1976; Thompson and Grime, 1978) the phenomenon is not restricted to such conditions. In vegetation subjected to summer drought, persistent seed banks are accumulated by both perennial herbs (e.g. *Silene nutans*, *Thymus drucei*, *Campanula rotundifolia*) and certain winter annuals (e.g. *Arabidopsis thaliana*, *Arenaria serpyllifolia*, *Saxifraga tridactylites*) (King, 1976; Thompson and Grime, 1978).

Regeneration from banks of persistent buried seeds in shrubs and trees is restricted to species which are associated with frequently-disturbed vegetation. In second-growth tropical forest seed banks of woody species are a common feature (Blum, 1968; Gómez-Pompa, 1967; Webb *et al*., 1972) whilst in temperate deciduous forest in North America this form of regeneration has been described in the pin cherry (*Prunus pensylvanica*). From his investigations with this species Marks (1974) concluded that 'sufficient numbers of viable pin cherry seeds reside in the soils of second-growth forests in central New Hampshire to account for the dense stands frequently observed after cutting or burning.'

An additional and rather unusual example of seed bank regeneration in woody species is provided by the closed-cone pines of California (*Pinus attenuata*, *P. contorta*, *P. muricata*, *P. radiata*, and *P. remorata*) and the Appalachian Mountains (*P. pungens*). From the results of several investigations (Munz, 1959; Zobel, 1969; Vogl, 1973) it is clear that the seed bank is specifically adapted for regeneration after periodic destruction by fire. In these relatively short-lived trees which occur in even-aged stands dating back to the most recent fire, the large cones remain firmly attached to the parent and accumulate throughout the life of each tree. Release of seeds is dependent upon the opening of the cones by heat generated during forest fires.

From this brief survey, it is apparent that persistent seed banks are associated with an extremely wide range of species and habitats. It is interesting to note, however, that large populations of buried seeds do not appear to accumulate in arctic regions or in areas occupied by mature tropical or temperate forest (Livingstone and Allessio, 1968). In these habitats, seed banks tend to be replaced by banks of persistent seedlings (see page 102).

The mechanism of a persistent bank of buried seeds

Although the functional significance of a bank of buried seeds varies in detail according to species and ecological situation, the same basic problems are encountered in any attempt to discover how the bank arises and is involved in seedling establishment. Firstly, it is necessary to describe the mechanism by which seeds become buried. Secondly, the factors must be identified which prevent the germination of seeds before burial and during their period of survival in the soil. Thirdly, since many seeds appear to germinate in places and at times propitious for seedling establishment, it is of considerable interest to examine the mechanisms which initiate the germination of buried seeds.

One approach to these problems has been to examine the morphology and germination physiology of seeds from species which develop persistent seed banks and to compare the results with those obtained for species with transient seed banks. From such comparisons it is possible to recognize certain characteristics which are consistently associated with the tendency to form a persistent seed bank. This suggests that for many plant species it may be possible to predict the presence of a seed bank from the laboratory characteristics of the seeds.

Seed burial In Table 12, a number of common British plants known to accumulate buried seed banks are listed, together with measurements of their average seed weight. A striking feature of this data is the high incidence of species with extremely small seeds. This applies particularly to species such as *Juncus effusus*, *Calluna vulgaris*, *Agrostis tenuis*, and *Origanum vulgare*, which are known to accumulate huge populations of buried seeds in their respective habitats. Small seeds are more likely than larger ones to be washed into small fissures in the soil surface and to be buried by the activities of the soil microfauna (McRill and Sagar, 1973; McRill, 1974) and evidence in support of this hypothesis has been obtained by Mortimer (1974) who observed the fate of marked seeds introduced into a grassland habitat.

It is interesting to note that some arable weeds such as *Avena fatua*, *Agropyron repens*, and *Polygonum convolvulus* are exceptional in that they have banks composed of comparatively large seeds. Here, however, it seems likely that burial occurs mainly as a result of ploughing of the soil.

The probability that a seed will become buried is likely to be increased by any dormancy mechanisms which delay germination in the period immediately after seedfall. Following the terminology of Harper (1957), we may distinguish two mechanisms of seed dormancy (innate and enforced), both of which appear to facilitate seed burial.

Table 12. Some flowering plants which, in Britain, have seeds which are partially or totally inhibited by darkness and tend to accumulate seed-banks in the soil. Mean seed weights are indicated in brackets (Thompson 1977).

(a) Herbaceous plants of grassland, heathland and woodland

Arabidopsis thaliana	(0.02)	*Origanum vulgare*	(0.10)
Arenaria serpyllifolia	(0.06)	*Poa trivialis*	(0.09)
Agrostis canina ssp. *montana*	(0.05)	*Potentilla erecta*	(0.58)
Agrostis tenuis	(0.06)	*Sagina procumbens*	(0.02)
Anthoxanthum odoratum	(0.45)	*Saxifraga tridactylites*	(0.01)
Calluna vulgaris	(0.03)	*Silene nutans*	(0.27)
Campanula rotundifolia	(0.07)	*Sieglingia decumbens*	(0.87)
Digitalis purpurea	(0.07)	*Thymus drucei*	(0.11)
Holcus lanatus	(0.32)	*Urtica dioica*	(0.20)
Milium effusum	(1.20)		

(b) Marshland plants

Agrostis canina ssp. *canina*	(0.05)	*Juncus articulatus*	(0.02)
Cirsium palustre	(2.00)	*Juncus effusus*	(0.01)
Deschampsia cespitosa	(0.31)	*Juncus inflexus*	(0.03)
Epilobium hirsutum	(0.05)	*Ranunculus sceleratus*	(0.08)
		Rumex sanguineus	(1.13)

(c) Arable weeds

Matricaria matricariodes	(0.08)	*Poa annua*	(0.26)
Polygonum aviculare	(1.45)	*Stellaria media*	(0.35)

Examples of innate dormancy include the need for an extended period of incubation in warm moist conditions for maturation of the embryo (e.g. *Ranunculus repens*, *Ranunculus acris*, and *Potentilla erecta*), inhibition of germination by light (e.g. *Agropyron repens*, *Cirsium arvense*, *Poa pratensis*), chilling requirements (e.g. *Polygonum aviculare*, *Galium palustre*) and the presence of an impermeable testa (e.g. *Helianthemum chamaecistus* and the majority of Leguminous herbs).

Enforced dormancy occurs in circumstances where seeds are prevented from germination by the conditions prevailing in the habitat during the period following seedfall. In some British species (e.g. *Calluna vulgaris*, *Rumex sanguineus*, *Urtica dioica*) the release of the seeds from the parent plant is so delayed, and the temperature requirements for germination are so high, that it is inevitable that dormancy will be enforced by low winter temperatures.

A quite different mechanism of enforced dormancy may operate in habitats in which imbibed seeds on the soil surface experience light which has been filtered by a dense leaf canopy. Laboratory experiments conducted by Taylorson and Borthwick (1969), King (1975), Grime and Jarvis (1975), and Gorski (1975) suggest that modification of the spectral composition of the incident radiation may be sufficient to enforce dormancy in certain species (Table 13), and from data such as those in Table 14 there is some evidence that germination of the

Table 13. The effect of light of different wavebands upon the germination of seeds of two groups of herbaceous plants differing in their capacity to form persistent seed banks in the soil (Grime & Jarvis, unpublished). Daylength: 16 h (20°C); nightlength: 8h (15°C). Mean seed weights, for the populations tested, are also included.

Species	Seed weight (mg)	Low intensity (6μ W cm^{-2})					High intensity (28μ W cm^{-2})				
		Dark	Far-red	Red	Blue	Green	Dark	Far-red	Red	Blue	Green
(a) Herbs without persistent seed banks											
Aira caryophyllea	0.09	94	92	91	90	90	83	0	88	55	93
Aira praecox	0.18	98	86	92	76	90	98	0	96	86	93
Bromus erectus	4.81	58	60	52	58	76	67	0	71	54	69
Bromus mollis	3.15	88	86	94	94	96	84	0	84	86	90
Bromus sterilis	8.37	82	90	85	86	85	87	0	74	96	90
Catapodium rigidum	0.19	82	76	94	89	96	81	0	95	81	97
Hordeum murinum	6.55	94	96	96	98	96	98	42	98	84	98
Lolium perenne	1.79	76	78	80	77	91	86	18	76	86	96
(b) Herbs with persistent seed banks											
Agrostis tenuis	0.06	64	69	93	52	74	55	3	69	34	82
Arabidopsis thaliana	0.02	5	1	56	3	27	0	1	53	10	21
Campanula rotundifolia	0.07	0	9	46	21	46	3	0	68	4	32
Chenopodium rubrum	0.09	0	1	0	0	0	0	1	62	0	12
Digitalis purpurea	0.07	2	49	72	35	71	0	0	66	11	64
Epilobium hirsutum	0.05	0	0	80	0	8	0	0	96	2	68
Juncus articulatus	0.02	0	0	2	0	0	0	0	42	0	0
Juncus effusus	0.02	0	0	2	0	0	0	0	46	0	0
Rumex crispus	1.33	3	2	90	7	40	4	4	92	12	93
Rumex obtusifolius	1.10	1	1	75	1	10	1	0	59	1	63
Sagina procumbens	0.02	0	43	90	28	72	0	0	97	32	96
Sieglingia decumbens	0.87	51	75	95	30	48	41	0	85	24	97

Table 14. The influence of a leaf canopy upon the percentage germination of seeds of two groups of herbaceous plants of contrasted ecology (Fenner 1978).

		Full sunlight		Shaded by *Festuca rubra*	
		Filter paper in closed petri dish	Bare soil in plant pot	Short turf in plant pot	Tall turf in plant pot
Species of open habitats	*Inula conyza*	94	70	72	16
	Reseda luteola	100	52	57	30
	Sonchus oleraceous	98	27	36	12
	Spergula arvensis	96	44	35	10
	Verbascum thapsus	76	45	45	27
Closed-turf species	*Galium verum*	70	56	57	39
	Hypochoeris radicata	98	80	62	29
	Leontodon hispidus	100	81	79	59
	Plantago media	88	36	65	38
	Rumex acetosa	100	75	75	42
	Scabiosa columbaria	92	60	66	40

seeds of ruderal herbs is more strongly inhibited by a dense leaf canopy than that of closed-turf species.

Dormancy in buried seeds In many of the species which develop persistent seed banks it is necessary to draw a distinction between the mechanisms which inhibit germination prior to burial and those which become important once the seed is incorporated into the soil. Some authorities such as Bibbey (1948) and Harper (1957) have contended that the composition of the soil atmosphere is a major factor restricting the germination of buried seeds. However, as the results of large-scale comparative experiments on the germination requirements of native species have become available (Table 13), it has become clear that a most consistent feature of the species forming persistent seed banks is the inhibition of germination by darkness. It seems reasonable to suspect, therefore, that in the majority of buried seeds dormancy is enforced primarily by lack of light. It is interesting to note, moreover, that it has been suggested (Wesson and Wareing, 1969b) that in some seeds (e.g. *Plantago lanceolata*) the requirement for light is not apparent immediately after the seeds are shed from the plant, but is induced later by a period of burial in the soil.

Initiation of germination in buried seeds It is well known that cultivation of agricultural soils usually results in the appearance of large numbers of seedlings of annual weeds, and it seems likely that this is at least in part due to the unearthing of buried seeds and the triggering of germination by exposure to light. The same explanation may be applied to the frequent observation that gaps in pastures caused by local damage to the turf are often rapidly colonized by seedlings. However, scrutiny of the seedlings which appear in response to disturbance of the soil (Thompson, 1977) has revealed that in some situations a high proportion of the seedlings originate from seeds which are situated beneath the soil surface below the depth to which light can penetrate. From investigations with naturally-buried seeds (Figure 24) and seeds placed in incubators simulating the temperature regimes experienced by seeds in the soil (Figure 27), there is evidence that the germination of buried seeds *in situ* is often brought about by a response to the increased diurnal fluctuations in temperature which may result from the removal of the insulating effect of foliage, litter, or humus layers from the soil

Figures 25 and 26 provide illustrations of the alteration of soil temperature regime which may result from the creation of gaps in a herbaceous canopy. From such data, it is apparent that the effect of a gap upon the diurnal fluctuations of temperature experienced by a buried seed is strongly affected by the depth of burial and the size of the gap. We may suspect therefore that the capacity to respond to particular amplitudes of temperature fluctuation in darkness acts as a depth-sensing mechanism and may also cause the germination of buried seeds to be restricted to gaps of a particular range in size.

Circumstantial evidence of the importance of sensitivity to fluctuating temperatures as a mechanism whereby gaps in vegetation may be detected by buried

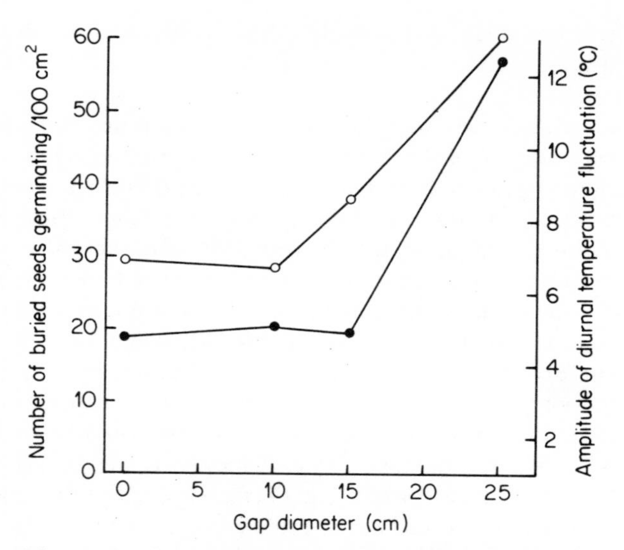

Figure 24 Comparison of soil temperature fluctuations (○) and germination (●) of naturally-buried seeds of *Holcus lanatus* in canopy gaps of different diameters in a sown pasture of low productivity in Northern England. Temperature fluctuations are based upon the mean amplitudes of diurnal fluctuation measured in the centre of circular gaps at a soil depth of 1 cm over the period 11–23 May 1977. Germination counts were carried out during the period 6 April–30 May (Thompson, 1977).

seeds is available from experiments (Figure 27) in which the capacity to respond to fluctuations in temperature in darkness has been found to occur in grasses such as *Poa annua*, *Poa trivialis*, *Poa pratensis*, *Holcus lanatus*, and *Deschampsia cespitosa*, all of which are prominent in the buried seed populations of pastures (Chippendale and Milton, 1934) and frequently exploit gaps in turf arising from excessive trampling and poaching by domestic animals.

Whilst the ability to respond to fluctuating temperatures in darkness is of widespread occurrence among species forming persistent seed banks (Thompson and Grime, 1978) this potential is consistently absent in species with extremely small seeds, all of which appear to have an obligatory light requirement. The significance of this phenomenon may be related to the fact that seedlings originating from small seeds are likely to be severely limited in their capacity to penetrate to the soil surface.

A feature of many of the herbaceous species with persistent seed banks is the tendency to exhibit polymorphism with respect to germination requirements. Seeds produced by the same population of plants (New, 1958; Grouzis *et al.*, 1976) or even one individual (Cavers and Harper, 1966) may show considerable differences in response to such factors as light and temperature and there may be extremely wide variation with respect to the amplitude of diurnal fluctuation

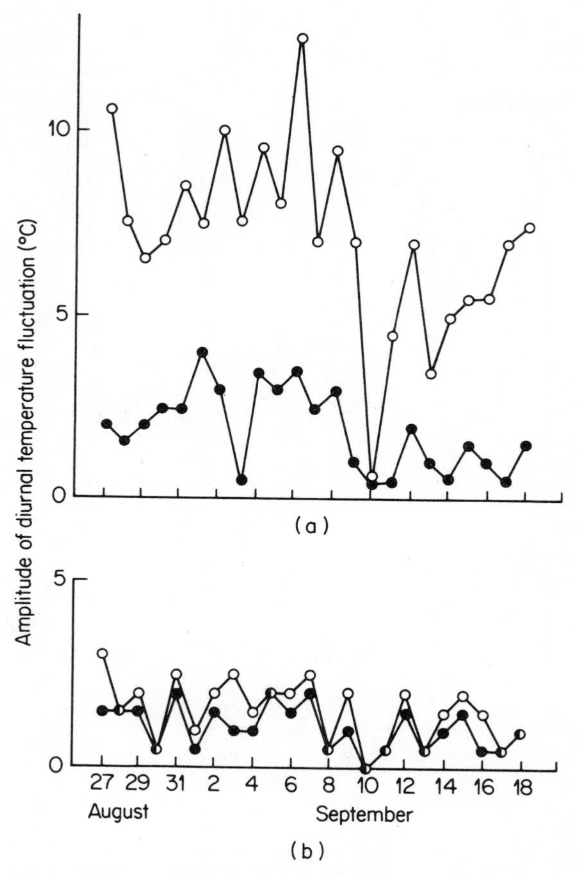

Figure 25 The amplitude of diurnal temperature fluctuations in soil temperature recorded beneath a dense leaf canopy mainly composed of the grasses *Dactylis glomerata* and *Arrhenatherum elatius* (●–●), and beneath artificially created gaps of 20 cm diameter (○–○), (a) at a soil depth of 1 cm, (b) at a soil depth of 5 cm. Measurements conducted in Sheffield, England, over the period 27 August–18 September 1977 (Thompson, 1977).

in temperature required to initiate germination (Figures 24 and 27). The significance of this 'polymorphism'* is particularly obvious in the ruderal species. Because these plants occupy disturbed environments, the risks of seedling mortality are high and synchronous germination of the seed bank could cause the

*Considerable caution should be exercised before attributing a genetic basis to this 'polymorphism' since there is now evidence to show that the germination physiology of a seed crop can be caused to vary by changing the environmental conditions experienced by the parent plant (Harrington and Thompson, 1952; Koller, 1962; Evenari, 1965; Karssen, 1968; Junttila, 1973; Guttermann, 1974).

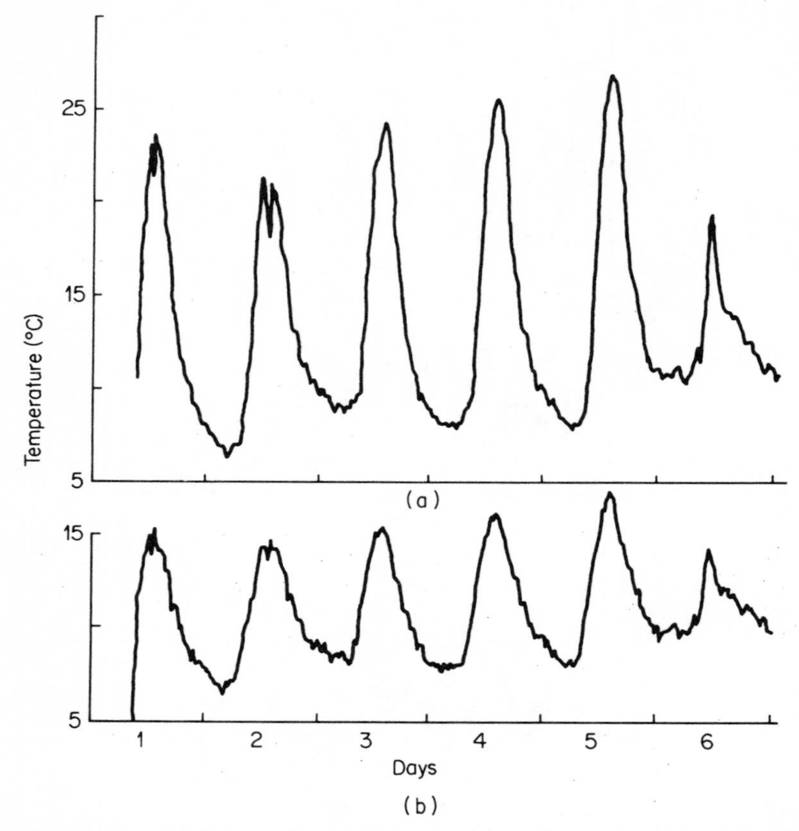

Figure 26 Diurnal changes in temperature at a depth of 1 cm in the soil in a sown pasture, mainly composed of *Lolium perenne*, *Holcus lanatus*, and *Agrostis stolonifera*, (a) beneath 25 cm diameter gaps, (b) beneath intact canopy (Thompson, 1977).

population to be vulnerable to extinction. It would appear therefore that in this context polymorphism not only provides the possibility of regeneration in a range of different spatial and temporal niches but also ensures the persistence of a reservoir of germinable seeds in the soil.

Regeneration involving numerous wind-dispersed seeds or spores (*W*)

In landscapes subject to erratic and large-scale disturbance (e.g., erosion, fire, tree felling, mining, quarrying, and road construction) it is not unusual for the flora to contain a high proportion of species which produce numerous wind-dispersed seeds. In temperate latitudes herbaceous plants of this type are particularly common in the *Compositae* (e.g. *Cirsium*, *Taraxacum*, *Crepis*, *Tussilago*, *Senecio*, *Petasites*), while shrubs and trees of the genus *Salix* also fall into this

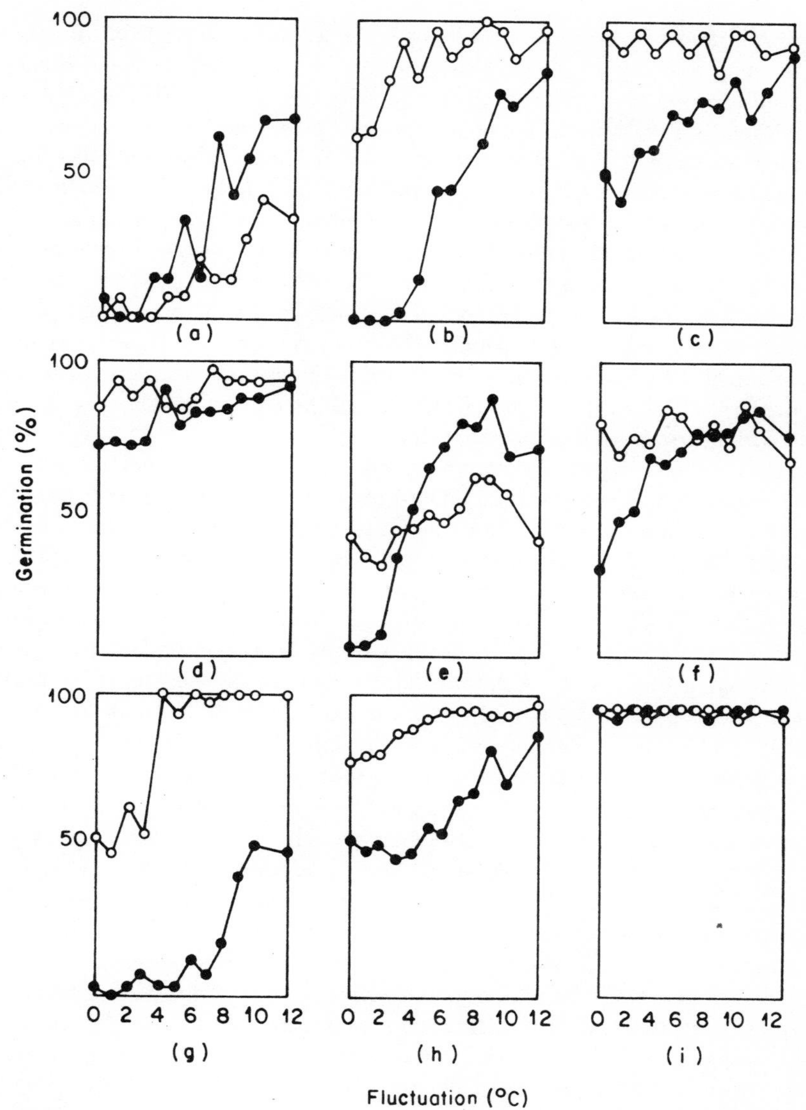

Figure 27 Germination responses to various amplitudes of diurnal fluctuations of temperature in light (o) and in continuous darkness (•) in nine herbaceous plants of common occurrence in pastures and arable land. Fluctuating temperatures were applied in the form of a depression below a base temperature of 22°C extending for 6 in every 24 h. Diurnal fluctuations were maintained throughout each experiment. Light intensity during the light period was 1846 μW cm^{-2}. Prior to testing, seeds were stored dry at 5°C. (a) *Agropyron repens,* (b) *Deschampsia cespitosa,* (c) *Holcus lanatus,* (d) *Poa annua,* (e) *Poa pratensis,* (f)*Poa trivialis,* (g) *Rumex obtusifolius,* (h) *Stellaria media,* (i) *Lolium perenne* (Thompson, Grime, and Mason, 1977). (Reproduced by permission of Macmillan (Journals) Ltd.)

category. Under favourable circumstances the fecundity of some of these plants may be such that each year extensive areas are 'saturated' by wind-dispersed seed, and each patch of bare ground is rapidly colonized by dense populations of seedlings. (The genetic consequences of 'saturating' invasions by wind-dispersed seeds are discussed on page 118). Such rapid colonization by *Tussilago farfara* has been described by Bakker (1960) for newly-drained polders in Holland, and in the British Isles a similar phenomenon is associated with the exploitation of disturbed ground by *Chamaenerion angustifolium*. Regeneration by means of small wind-dispersed seeds is also well-documented for the North American prairies where it is associated particularly with the initial phase of colonization of earth mounds produced by burrowing animals such as prairie dogs (Koford, 1958) and badgers (Platt, 1975). It seems reasonable to conclude that primarily this form of regeneration is adapted to exploit forms of disturbance which are spatially unpredictable.

Regeneration involving the copious production of wind-dispersed seeds also occurs among large, comparatively fast-growing trees, which in temperate regions are to be found in genera such as *Fraxinus*, *Acer*, *Ulmus*, *Ailanthus*, *Liriodendron*, and *Populus*. In aboriginal tropical forests this regenerative strategy is restricted to the so-called 'emergents' or 'nomads' (Keay, 1957; van Steenis, 1958; Knight, 1975), tall fast-growing trees which appear to be adapted to colonize large gaps created in the forest canopy by windfalls (see page 181).

A detailed classification of the mechanisms of wind-dispersal is beyond the scope of this book and is available from the works of Ridley (1930) and van der Pijl (1972). However, for ecological purposes some broad functional classes may be recognized. Table 15 contains a list of some common plants of the British flora which regenerate by means of numerous wind-dispersed seeds. These can be conveniently subdivided into three classes which differ with respect to the mechanism of dispersal. The first category includes a number of ferns and orchids in which the seeds (or spores in the pteridophytes) are exceedingly small and are capable of long-distance transport in air currents. In the ferns, it would seem likely that this form of dispersal resembles that of many bryophytes and lichens, in that it facilitates colonization of habitats inaccessible and/or inhospitable to the majority of flowering plants (crevices in cliffs, rocks, tree trunks, and walls). Such habitats present a small target for a colonizing species but once located they constitute 'safe-sites' (Sagar and Harper, 1961) amenable even to establishment by dispersules with such very small reserves. However, some of the orchids, with 'dust' seeds, have a different type of ecology and are components of grassland and other types of herbaceous vegetation. Here, as pointed out by Harper (1977), the sacrifice of seed reserve is consistent with the peculiar process of seedling establishment which in these orchids involves a dependent subterranean phase with a mycorrhizal associate. Natural selection for high fecundity and dispersal efficiency in such plants may be interpreted, therefore, as a mechanism increasing the chance of effective contact in suitable environments between orchid seed and fungal symbiont.

Table 15. Some plants which regenerate by means of numerous wind-dispersed seeds or spores and are of common, widespread or expanding distribution in the British Isles. The values in brackets are the mean weights of dispersules in populations sampled in Northern England. All the species in category (a) have mean seed or spore weights less than 0.01 mg.

(a) Species with "dust" seeds or spores			
Asplenium ruta-muraria		*Dryopteris filix-mas*	
Asplenium trichomanes		*Equisetum arvense*	
Athyrium filix-femina		*Lycopodium clavatum*	
Dactylorhiza fuchsii		*Orchis mascula*	
Dryopteris dilatata		*Pteridium aquilinum*	
(b) Species with plume or pappus			
Chamaenerion angustifolium	(0.05)	*Leontodon hispidus*	(0.85)
Cirsium vulgare	(2.64)	*Mycelis muralis*	(0.34)
Crepis capillaris	(0.21)	*Petasites hybridus*	(0.26)
Epilobium hirsutum	(0.05)	*Phragmites communis*	(0.12)
Epilobium montanum	(0.13)	*Salix cinerea*	(0.09)
Epilobium nerterioides	(0.02)	*Senecio squalidus*	(0.21)
Eriophorum angustifolium	(0.44)	*Taraxacum officinale*	(0.64)
Hieracium pilosella	(0.15)	*Tussilago farfara*	(0.26)
Hypochoeris radicata	(0.96)	*Typha latifolia*	(0.03)
Leontodon autumnalis	(0.70)		
(c) Species with winged seeds			
Acer pseudoplatanus	(34.47)	*Linaria vulgaris*	(0.14)
Alnus glutinosa	(1.30)	*Pinus sylvestris*	(5.59)
Betula pubescens	(0.12)	*Rhinanthus minor*	(2.84)
Fraxinus excelsior	(46.19)	*Ulmus glabra*	(9.93)

The second class of species in Table 15 consists of herbaceous plants in which dispersal of a larger seed is facilitated by structures such as a plume or pappus which increases the buoyancy of the dispersule. An interesting example of this phenomenon is provided by the composites *Tussilago farfara* and *Petasites hybridus* in which extremely rapid germination occurs over a wide range of temperatures in both light and darkness, the seeds being very short-lived. The seeds of these two plants are released during the early spring and their germination physiology appears to be specifically adapted for the rapid colonization of bare soil created during the winter by flooding of river terraces and disturbance of glacial moraines.

Many of the other species cited in the second category in Table 15 resemble *T. farfara* and *P. hybridus*, in that the seeds germinate rapidly over a wide temperature range, decline in viability rather rapidly, and do not appear to be incorporated into persistent seed banks. This does not mean that germination in all of these species invariably occurs immediately after contact of the seed with the soil surface. It is interesting to observe that, in Table 14, the wind-dispersed composite, *Hypochoeris radicata*, is among the species found by Fenner (1978) to be inhibited from germination in light filtered by a dense leaf canopy. This suggests that in at least some of the species in this category there may be mechanisms which by temporary delay of germination optimize the timing of germination by responding to seasonal changes in the canopy of established plants.

A more elaborate post-dispersal sequence may be suspected in a number of wind-dispersed marsh-plants (e.g. *Epilobium hirsutum*, *Typha latifolia*) and ruderals (e.g. *Sonchus asper*, *Crepis capillaris*, *Conyza canadensis*, *Senecio viscosus*) in which germination is inhibited by darkness (Grime, Rodman, and Lee, unpublished). This finding suggests that, in such species, effective wind-dispersal may be allied to the capacity of the seed to remain dormant in the soil. Studies of the composition of natural seed banks provide confirmation that, particularly in the case of marshland habitats, some wind-dispersed species such as *Typha glauca* (van der Valk and Davis, 1976) and *Epilobium hirsutum* (Thompson, 1977) may form a persistent and major component of buried floras. Further research is required in order to determine how commonly this potentially powerful combination of attributes occurs in nature.

The third class of wind-dispersed species in Table 15 is made up of plants in which seed buoyancy depends upon planar extensions of the fruit wall into membranes or wings. Perhaps the most interesting members of this group are trees such as *Acer pseudoplatanus* and *Fraxinus excelsior* in which the rather large winged seeds have only a modest potential for wind-dispersal and depend heavily upon the fact that many of the seeds are released from a considerable height during windy conditions. It is interesting to consider why natural selection in these species has failed to produce smaller seeds such as those of *Betula*, capable of dispersal over greater distances. The explanation seems to be that dispersule size in these trees represents a compromise between the requirement for short-distance dispersal into clearings or beyond expanding forest margins and the need for an embryo and seed reserve large enough to facilitate establishment in herbaceous vegetation (see page 139).

Regeneration involving a bank of persistent seedlings ($B\Upsilon$)

In the majority of forest trees there is a comparatively short interval between seed-drop and germination, and in tropical species germination is often extremely rapid (van der Pijl 1972). Among temperate forest trees it is not uncommon for either germination itself (e.g. *Fagus sylvatica*) or plumule extension (e.g. *Quercus petraea*) to be temporarily delayed, but there is no evidence that such species accumulate persistent seed banks. However, for some considerable time it has been recognized (Brown, 1919; Sernander, 1936) that in mature temperate and tropical forests a very common regenerative strategy is that in

Table 16. A list of tree species regenerating by means of persistent seedlings.

Species	Source of information	
	Country	Observers
Tsuga canadensis (Hemlock)	North America	Marshall (1927)
Picea abies (Norway spruce)	Sweden	Sernander (1936)
Quercus alba (White oak)	North America	Oosting & Kramer (1946)
Ilex aquifolium (Holly)	England	Peterken & Lloyd (1967)
Acer saccharum (Sugar maple)	North America	Keever (1973)
Fagus grandifolia (Beech)	North America	Marks (1974)

which populations of tree seedlings and saplings persist for long periods in an extremely stunted or etiolated condition and a small percentage survive to maturity by expanding into canopy gaps when these arise through the senescence and death of established trees. In Table 16, a list is presented of trees for which regeneration by such 'advance reproduction' (Marks, 1974) has been documented.

In the population dynamics of these trees, the reservoir of seedlings and saplings functions in a way which is in some respects analogous to that of a seed bank. The similarity extends even to the critical role of disturbance of the established vegetation by senescence, pathogenic fungi, and windfalls in releasing individuals from the bank.

In many forest trees, seeds are not produced each year and the capacity of the seedlings to survive for long periods under unfavourable circumstances ensures that the potential for regeneration is maintained. A factor which may contribute to the persistence of the seedlings of many trees in temperate and tropical forest is the size of the seed (Salisbury, 1942), which in many species is exceptionally large in relation to the rate of seedling growth and provides an effective buffer during initial establishment in heavily-shaded and/or nutrient-deficient environments (Grime and Jeffrey, 1965).

In forests, reproduction by means of slow-growing seedlings is not confined to large-seeded trees. In his pioneer studies of plant demography, Tamm (1956) recorded the presence of a bank of extremely persistent slow-growing seedlings in a population of the shade-tolerant evergreen herb *Sanicula europaea* and from several laboratory investigations (Bordeau and Laverick, 1958; Hutchinson, 1967; Loach, 1970; Mahmoud and Grime, 1974) there is abundant evidence of the role of persistent, slow-growing seedlings in the regeneration of shade-plants. The results of a number of field observations and experiments suggest that the same regenerative strategy is also characteristic of arctic—alpine species (Billings and Mooney, 1968) and plants of nutrient-deficient soils (Kruckeberg, 1954; Beadle, 1954; Rorison, 1960; Davison, 1964; Rabotnov, 1969; Grime and Curtis, 1976).

THE ROLE OF ANIMALS IN REGENERATION

In the description of the mechanism of a persistent seed bank (page 91) reference was made to the involvement of animals both in the burial of seeds in the soil and in the initiation of germination in buried seeds. Animals also exert two other major impacts upon regeneration. One is the result of the predation of seeds and seedlings by animals (see page 111) whilst the second concerns the dispersal of seeds or vegetative propagules.

The mechanisms whereby animals may affect the dispersal of seeds are many and various. From work such as that of Ridley (1930) and van der Pijl (1972) considerable progress has been made in the task of describing and classifying seeds and fruits which are dispersed by animals. It is evident that in a large number of plants there are adaptations which facilitate transport of the seed by particular animals, and a broad difference can be recognized between dispersal mechanisms (e.g. burrs, hooked fruits, glutinous seeds) which involve attachment to the exterior of the animal (ectozoochory) and those in which the dispersule is attractive and all or part of it is eaten by vertebrate or invertebrate animals (endozoochory). Within this second category a further distinction can be drawn between certain plants (e.g. *Chenopodium rubrum*, *Vaccinium myrtillus*, *Polygonum persicaria*, *Urtica dioica*) in which some or all of the seeds are capable of surviving passage through the gut of various animals and others, including many large-seeded forest trees (e.g. *Quercus* spp., *Castanea* spp., *Carya* spp., *Fagus* spp.), in which the seeds are readily digested and successful dispersal and establishment depends upon the small proportion of seeds which are 'lost in transit' or escape predation in 'unclaimed caches' (Reynolds, 1958; Van der Wall and Balda, 1977).

In the context of a general description of the main types of regenerative strategies a detailed analysis of the role of animals in seed dispersal need not be attempted. However, it is necessary to point out that adaptation for seed dispersal by animals is frequently associated with three of the five regenerative strategies described in this chapter. From the examples cited in Table 17, it is clear that zoochory frequently occurs as a prelude to seasonal regeneration and may precede incorporation into a buried seed bank or a bank of slow-growing seedlings.

MULTIPLE REGENERATION

A complication of major significance arises from the fact that several regenerative strategies may be exhibited by the same population or genotype. This phenomenon has important consequences for both the ecological and evolutionary potential of the plant. (The evolutionary implications of multiple regeneration are examined on pages 116–118.)

In order to illustrate some of the combinations of regeneration strategies which may occur, a number of examples referring to populations examined in

Table 17. Examples of the role of animals as dispersal agents in three types of regenerative strategies.

		Plant species	Dispersal agent	Authority
(a)	Regenerating by seasonal colonization of vegetation gaps	*Hordeum* spp.	Various agents	Stebbins (1971)
(b)	Regenerating from a persistent seed bank	*Pedicularis sylvatica*	Ants	Berg (1954)
		Polygonum lapathifolium *Polygonum persicaria*	Cottontail rabbit (*Sylvilagus floridanus*)	Stainforth & Cavers (1977)
		Prunus pensylvanica	Various birds	Olmsted & Curtis (1947)
(c)	Regenerating from a bank of persistent seedlings	*Casearia corymbosa*	Masked Tityra (*Tityra semifasciata*)	Howe (1977)
		Pinus edulis	Clark's Nutcracker (*Nucifraga columbria*)	van der Wall & Balda (1977)
		Calvaria major	Dodo (*Raphus cucullatus*)	Temple (1977)
		Ilex aquifolium	Various birds	Peterken & Lloyd (1967)

Table 18. The regenerative strategies exhibited by various flowering plants of common occurrence in the British Isles.

Species	Regenerative strategies				
	Vegetative expansion	Seasonal regeneration	Persistent seeds	Numerous wind-dispersed seeds	Persistent seedlings
Impatiens glandulifera		*			
Dactylorrhiza fuchsii				*	
Quercus petraea					*
Salix cinerea	*			*	
Ilex aquifolium	*				*
Tussilago farfara		*		*	
Digitalis purpurea		*	*		
Bromus erectus	*	*			
Poa annua (ruderal population)		*	*		
Poa annua (pasture population)	*	*	*		
Agropyron repens	*	*	*		
Epilobium hirsutum	*	*	*	*	

Britain have been assembled in Table 18. Six of these plants will now be considered in a series which is one of increasing complexity with respect to mechanisms of regeneration.

Example 1. Impatiens glandulifera. References have been made already (pages 58 and 84) to this summer annual which regenerates exclusively by means of large seeds which germinate synchronously in the spring. The result of this restriction to a single regenerative strategy is that there is no capacity for vegetative expansion nor for the development of a buried seed bank. The ecological consequence is the confinement of populations to streamsides and to other habitats in which there is a major impact of seasonally-predictable disturbance.

Example 2. Digitalis purpurea. In this biennial or short-lived perennial regeneration is by means of small seeds which are produced in very large numbers at the end of the vegetative phase. Where seeds fall onto bare soil germination and seedling establishment can occur directly, and in habitats where there is extensive and continuous disturbance, such as trackways in commercial forests and plantations, very high densities of established plants develop. However, the seeds of *D. purpurea* are inhibited from germination by darkness and are capable of extended periods of dormancy after burial in the soil. It is not surprising therefore to find that in many woods, copses, and hedgerows small populations often persist in vegetation in which disturbance is infrequent and more localized.

Example 3. Poa annua (ruderal populations). In arable land, gardens, paths, lawns, and intensively-grazed pastures, this small grass is a major vegetation component. Reference to studies of the life-cycle and reproductive biology in disturbed habitats (Wells, 1974; Thompson, 1977; Law, Bradshaw, and Putwain, 1977) suggests that in this type of habitat *P. annua* is represented by populations in which regeneration is mainly by seed and, as in the case of *Digitalis purpurea*, incorporates two regenerative strategies, i.e. there is a large burst of seedling germination and establishment from freshly-shed seed but regeneration also involves subsequent recruitments from a persistent seed bank in the soil.

Example 4. Poa annua (pasture populations). In grassland populations of *P. annua* it is not unusual for a high proportion of individuals to be robust plants which are capable of expanding laterally into neighbouring gaps by means of tillers which root at the nodes. Experiments in which seed progeny of *P. annua* from pasture populations have been grown under standardized conditions (Law, Bradshaw, and Putwain, 1977) have revealed that flowering and seed production is considerably delayed in comparison with ruderal populations and there is a corresponding increase in longevity and vegetative vigour (Plate 13). It would appear, therefore, that under certain pasture conditions a third dimension has been added to the regenerative capacity of *P. annua* through the development of vegetative reproduction by rooting tillers.

Example 5. Agropyron repens. Couch-grass (*A. repens*) is a coarse, strongly-rhizomatous grass which is a persistent weed in arable land but is also of widespread occurrence in less disturbed habitats such as pastures, roadsides, and derelict land. Three forms of regeneration may be recognized in this plant. The first consists of vegetative expansion by means of the extensive rhizomatous system which allows the plant to form large patches in areas of productive herbaceous vegetation dominated by perennial species. The two remaining regenerative strategies of *A. repens* are particularly important in disturbed habitats. One involves regeneration from a persistent seed bank whilst the other depends upon vegetative propagation from pieces of rhizome fragmented and dispersed during cultivation of the soil.

Example 6. Epilobium hirsutum. Regeneration in this tall rhizomatous wetland herb is characterized by an extraordinary array of strategies. In relatively stable habitats, *E. hirsutum* is capable of rapid clonal spread by extensive rhizomes. The species also regenerates by means of numerous small wind-dispersed seeds, some of which germinate immediately whilst others are incorporated into a persistent seed bank (Thompson, 1977). In disturbed habitats *E. hirsutum* may also regenerate from rhizome fragments.

These few examples and those cited in Table 19 illustrate only a small proportion of the combinations of regenerative strategies which exist in natural populations. Certain combinations are characteristic of plants from particular habitats (see pages 112–115), and here perhaps the best example is the consistent association in stress-tolerant herbs and shrubs between vegetative expansion and regeneration involving a bank of persistent seedlings.

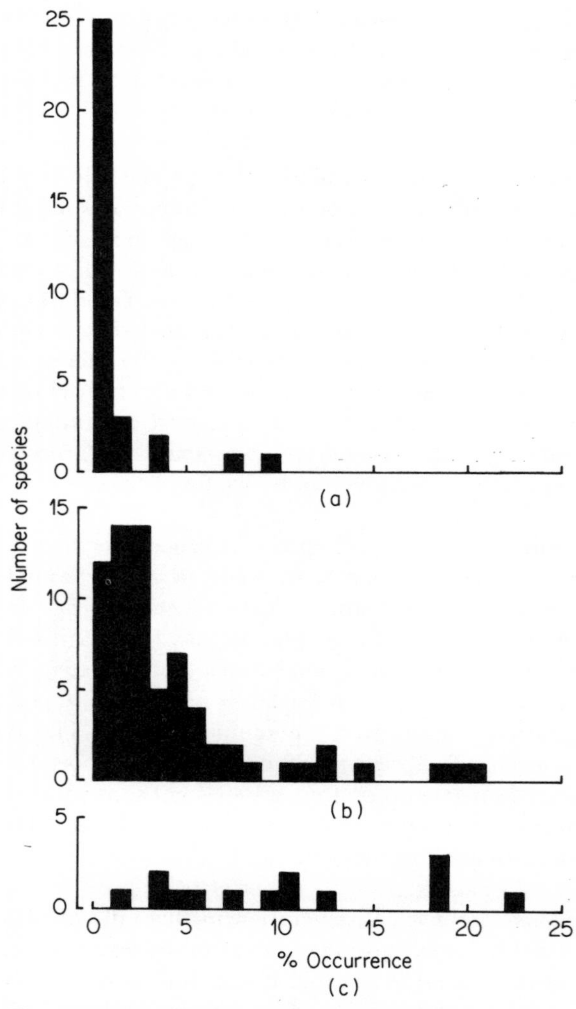

Figure 28 Histograms illustrating the frequency of occurrence of grassland plants in the Sheffield area: (a) species with a single regenerative strategy, (b) species with two regenerative strategies, (c) species with three or more regenerative strategies. Frequency is based upon the percentage occurrence of the species in 2748 m^2 quadrats distributed within the 32 major habitats occurring in an area of 2400 km^2.

We may predict that possession of several strategies will enlarge the range of circumstances in which regeneration can occur and this, in turn, may be expected to widen the ecological range of the plant and to confer a greater degree of persistence under fluctuating environmental conditions. The ecological range of a plant is also, of course, a function of characteristics manifested during the established phase of the life-history. Moreover, the abundance and ecological amplitude will be affected also by the degree of genetic variability within and between populations. Nevertheless, we may expect that in an environmentally heterogeneous landscape there will be a general relationship between the frequency of occurrence of a species or ecotype and the number of regenerative strategies which it employs. Evidence consistent with this hypothesis has been obtained from calculations based upon surveys of the grassland flora of the Sheffield region (Figure 28). From these data it is apparent that multiple regeneration is associated with a marked expansion in abundance and ecological amplitude.

REGENERATION FAILURE

From earlier sections of this chapter it is apparent that each type of regenerative strategy is adapted to a limited range of habitat conditions. We may predict, therefore, that the failure of a plant to regenerate outside this range will be a major determinant of its ecological amplitude and geographical distribution. In addition we may expect that, within the habitat range, there will be marked effects of physical and biotic factors upon regenerative capacity and these will modify the abundance and status of the plant in different types of vegetation. Moreover, in a plant which exhibits multiple regeneration, profound changes in ecology may be anticipated where the effect of environment is to alter the relative effectiveness of its various regenerative mechanisms.

Failure in plants with a single regenerative strategy

Annual and biennial herbs

The survival of populations of annual and biennial plants depends upon successful regeneration by seed, and in this process there are a number of stages (flower production, pollination, seed development, dispersal, germination, seedling establishment) where an unfavourable environment may intervene to prevent effective regeneration.

Reference has been made already (page 23) to the fact that annual plants are relatively scarce in arctic and alpine regions, and from experiments such as those of Davison (1977) with the annual *Hordeum murinum* it is evident that major limiting effects in these environments are the low temperatures and the short duration of the growing season, factors which severely reduce the level of seed production (Table 19) and which may completely inhibit regeneration in particularly unfavourable years.

Table 19. Comparison of flower and fruit production in the annual grass *Hordeum murinum* growing at contrasted altitudes in Northern England (Davison 1977).

Site	Altitude (m)	Mean number of inflorescences per plant	Mean number of fruits per plant
Newcastle	91	26.0	728
Moorhouse	558	9.8	235

Even in geographical regions where the climate allows high rates of seed production, annual and biennial plants are unable to colonize productive, relatively undisturbed herbaceous vegetation, and here the breakdown of the regenerative process usually occurs at the stage of seedling establishment. It is clear from earlier parts of this chapter that seedling establishment in many, if not all, plants is adversely affected by the presence of a dense cover of established plants and litter, and in a large number of annual and biennial species successful regeneration by seed occurs only by the exploitation of gaps. One consequence of this dependence is that, in grasslands, annual plants become most important in areas of low rainfall where the effect of desiccation is to restrict the development of a continuous cover of foliage and litter. At northern latitudes, grassland annuals are often restricted in occurrence to areas of shallow soil on south-facing slopes (Ratcliffe, 1961; Perring, 1959; Grime and Lloyd, 1973).

In Britain, unfertilized pastures on chalk or limestone contain a diverse assemblage of annual and biennial plants, many of which have declined in abundance during the last forty years (Perring and Walters, 1962; Perring, 1968). The disappearance of these plants from many localities has been associated with fertilizer applications and/or reduced grazing by sheep and rabbits, changes which have produced a more dense and uniform cover of foliage and litter and which, as pointed out by Grubb (1976), has obliterated the gaps upon which many of the annual and biennial species depend for seedling establishment.

Trees and shrubs

There are many trees and shrubs in which seedling establishment is the only effective method of regeneration and we may expect therefore that the ecology of woody species will be restricted by effects of environment upon seed production and seedling mortality. As in the case of the annuals and biennials, discussed under the last heading, seed production in many trees and shrubs is severely reduced at northern latitudes. The causes of this phenomenon vary with species and location but they include frost damage to flowers and the failure of fruits to reach maturity.

Within the regions in which seed production is relatively unrestricted by climate, a factor limiting the regeneration of many shrubs and trees is predation by animals and fungi. Seeds and fruits of trees are consumed by a wide range of animals and they are the primary food source in many birds, rodents, and primates. It is to be expected, therefore, that in addition to their incidental role as agents of dispersal, animals will often function as the most important factor controlling the proportion of the seed output which survives and germinates (Smith, 1970; Batzli and Pitelka, 1970; Gardner, 1977).

Predation also operates during the seedling phase and the results of experiments (Vaartaja, 1952; Vaartaja and Cran, 1956; Grime and Jeffrey, 1965) and observations under natural conditions (Wardle, 1959; Ross *et al.*, 1970; Janzen, 1970) suggest that fungal attack and selective feeding by small mammals and invertebrates provide an inconspicuous but potent influence on seedling survival.

Effects of predation by animals are likely to be particularly important in mature forest or scrub where regeneration often depends upon the maintenance of a bank of slow-growing seedlings and saplings. In many parts of the world, including Australia (Dixon, 1892; Crisp and Lange, 1976) and North America (Roughton, 1972), regeneration of woody species is restricted by domestic animals such as sheep and goats. This phenomenon is also well known in Britain where regeneration in many of the surviving fragments of unfenced northern oakwoods is prevented by the destructive effect of sheep grazing.

Failure in plants with more than one regenerative strategy

In perennial species with several mechanisms of regeneration it is not uncommon to find populations in which the effect of the environment is to reduce the number of 'effective' regenerative strategies to one. This phenomenon is frequently observed in populations which lie near to the limits of the geographical distribution of a plant species. One of the earlier documented examples of this phenomenon is that of Mooney and Billings (1961) who showed that, the further north their situation, populations of the herbaceous perennial *Oxyria digyna* show decreasing amounts of seed production and increased dependence upon vegetative reproduction. In Britain there are several perennial herbs, such as *Cirsium acaulon*, *Brachypodium pinnatum*, and *Filipendula vulgaris*, in which the most northerly populations rely almost exclusively upon vegetative expansion and only rarely regenerate by seed. After careful comparison of populations at contrasted latitudes over several years Pigott (1968) has concluded that the very low frequency of regeneration from seed by *Cirsium acaulon* at its most northerly stations is due to reductions in the number of mature seeds produced and more especially to infection by the fungus *Botrytris cinerea* which causes the seeds to rot whilst they are still attached to the parent plant.

It is not uncommon, as a consequence of partial regenerative failure, for populations of the same species in neighbouring habitats to depend upon different regenerative strategies. As we have seen in the case of *Poa annua* (page 107).

genetic differences between populations may be involved in this phenomenon. However, in many instances, the explanation lies in the fact that local differences in the physical environment or in vegetation management cause changes in the relative effectiveness of alternative forms of regeneration. For example, in many British woodland herbs and shrubs, e.g. *Deschampsia flexuosa*, *Lonicera periclymenum*, *Rubus fruticosus* agg., regeneration in shade is almost exclusively restricted to vegetative expansion, although in clearings nearby it is not unusual to find high rates of seed production. Similar contrasts abound in the British landscape where closely adjacent sites are often subjected to very different forms of management. In roadsides and disturbed habitats in suburban areas common grasses such as *Poa trivialis*, *Lolium perenne*, and *Holcus lanatus* frequently regenerate from seed, whereas in neighbouring lawns and pastures the same species may be almost exclusively dependent upon vegetative forms of regeneration.

A particularly interesting form of regeneration failure is commonly observed in dioecious plants where seed production may be inhibited by geographical isolation of the sexes. In Britain, two well-known examples of species illustrating this phenomenon are *Petasites hybridus* and *Mercurialis perennis*. These herbs are capable of forming extensive stands by vegetative expansion and, in both species, it is not uncommon for local populations to be composed exclusively of either male or female plants.

CONCLUSIONS

From the evidence which has been reviewed in this chapter it would appear that differences in the intensity, periodicity, and spatial distribution of disturbance have provided the major selective forces in the differentiation of regenerative strategies. This conclusion is consistent with the view that regenerative strategies have evolved primarily as mechanisms whereby the juvenile stages in life-histories tolerate or evade the potentially dominating effects of established plants.

Having established the basis upon which to classify regenerative strategies, it remains to identify more precisely their role in particular life-histories. Only when a great deal more is known about the biology of plant populations under natural conditions will it be possible to describe in detail the genetic and ecological consequences of various regenerative strategies. However, even at this early stage of data collection and interpretation it is apparent that the impact of a particular regenerative strategy is modified by two important variables. The first is the strategy adopted during the established phase of the life-cycle whilst the second is the breeding system of the plant.

Combinations between strategies in the established and regenerative phases of the life-cycle

In Figure 29 an attempt has been made to summarize in broad terms the range of associations occurring in flowering plants between particular regenera-

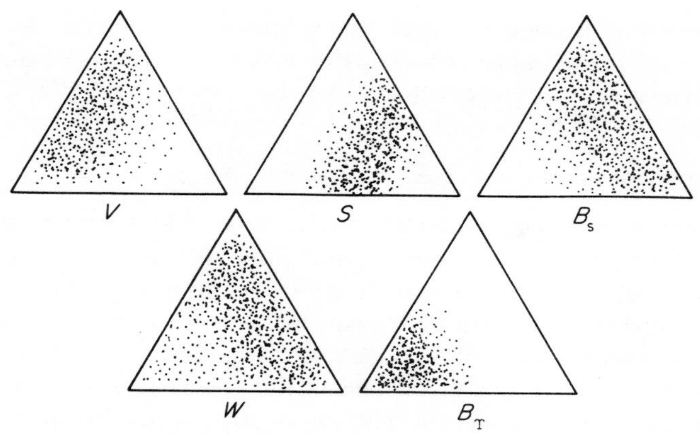

Figure 29 Model describing the range of combinations occurring in flowering plants between regenerative strategies and strategies of the established phase of the life-cycle: *V*, combinations involving vegetation expansion; *S*, combinations involving seasonal regeneration; B_s, combinations involving a persistent seedbank; *W*, combinations involving numerous wind-dispersed seeds or spores; B_{Υ}, combinations involving persistent seedlings. The distribution of established strategies within the triangular model is the same as that described in Figure 18.

tive strategies and strategies of the established phase. In this figure it is suggested that each of the five main regenerative strategies occurs in association with a particular range of established strategies. The overlap of distributions evident in this figure is consistent with the various types of multiple regeneration which occur in natural populations, including those cited in Table 18.

Combinations involving vegetative expansion

The outstanding feature of the distribution of vegetative expansion in the triangular model is its association with established strategies characteristic of relatively undisturbed habitats (competitors, stress-tolerant competitors, and stress-tolerators). This pattern appears to be related to the fact that vegetative expansion involves a period of attachment between parent and offspring and is not viable in circumstances in which the vegetation is affected by frequent and severe disturbance.

It seems likely that the role of vegetative expansion may be rather different in competitors and stress-tolerators. In productive, relatively undisturbed habitats, vegetative expansion is often an integral part of the mechanism whereby competitive herbs, shrubs, or trees rapidly monopolize the environment and suppress the growth and regeneration of neighbouring plants (see Chapter 4).

Under the very different conditions in severely stressed environments, however, the main advantage of vegetative expansion derives from the capacity to sustain the offspring under conditions in which establishment from an independent propagule is a lengthy and hazardous process.

Combinations involving seasonal regeneration in vegetation gaps

This regenerative strategy is particularly associated with competitive-ruderals and stress-tolerant ruderals, both of which, as explained on pages 58 and 63, are restricted to sites where the intensity of disturbance is sufficient to cause vegetation gaps to occur with high frequency and constancy. It would appear, therefore, that these two combinations arise from the fact that, in each case, the strategy in the regenerative phase and in the established phase is adapted to habitats subject to moderately-severe, seasonally-predictable disturbance.

Seasonal regeneration involves propagules which lack long-term dormancy, and in habitats subjected to more severe and/or less predictable forms of disturbance we may expect that this attribute will limit the regenerative capacity. It is probably for this reason that the incidence of exclusively seasonal forms of regeneration is low among ruderals *sensu stricto*.

Combinations involving a persistent seed bank

Regeneration involving a persistent seed bank is commonly associated with all of the established strategies with the exception of the stress-tolerators where the seed bank is usually replaced by a bank of persistent seedlings.

Although some of the plants which develop persistent seed banks have mechanisms which facilitate seed dispersal by animals (Table 17), most seed banks are located in close proximity to the parent plants and may even be situated directly beneath them or, in the extreme example of fire-adapted trees (page 90), they may actually remain attached to the parent. It seems likely, therefore, that the main functional significance of a seed bank is related to the extent to which it allows regeneration *in situ*. This interpretation is consistent with the fact that seed banks are particularly common in proclimax vegetation such as grasslands, heathland, and disturbed marshes, in which the process of vegetation change is cyclical rather than successional. Many forms of proclimax vegetation, especially those in farmland, are subject to alternating patterns of land-use with the result that in each area conditions suitable for particular established plants are available intermittently. In this situation the presence in the soil of a persistent seed bank allows the survival of populations during periods in which the management regime is unfavourable to the established plant.

Persistent seed banks occur in association with a wide range of primary and secondary strategies and it is clear that in certain respects the role of a seed bank changes according to the established strategy with which it is associated. In ruderals such as the annuals of arable fields and marshland the seed bank allows rapid recovery from catastrophic mortalities inflicted by cultivation, herbicide treatments, or natural phenomena such as flooding, and also permits survival of periods of temporary stability during which the ruderals are excluded by perennial species. In competitive-ruderals and stress-tolerant ruderals and C–S–R strategists the role of the seed bank is similar to that which it exerts in the ruderal except that here it appears to be mainly concerned with regeneration within gaps in perennial vegetation. In habitats dominated by competitors and stress-tolerant competitors, the intervals between major disturbances may be very long (>15 years). Competitors and stress-tolerant competitors with persistent seed banks include many of the most familiar herbs and shrubs of vegetation types subject to occasional disturbance by factors such as burning or flooding. After destruction of heathland vegetation by fire, for example, the large seed-banks of *Calluna vulgaris* enable this species to recover its dominant status relatively quickly (Whittaker and Gimingham, 1962; Hansen, 1976) and a similar phenomenon is involved in the rehabilitation of *Juncus effusus* and *Urtica dioica* in disturbed marshes and cultivated ground respectively.

Combinations involving numerous small wind-dispersed seeds or spores

Combinations of this type involve a range of established strategies similar to that described for buried seed banks but, in addition, they occur in association with some stress-tolerators including lichens and certain bryophytes, ferns, and angiospermous epiphytes.

There are major differences in ecology between species and populations which produce buried seed banks and plants which regenerate by means of numerous small wind-dispersed seeds or spores. The latter appear to be fugitives (Hutchinson, 1951) adapted to exploit landscapes subject to spatially unpredictable disturbance. Where disturbance occurs as an isolated and unusual event in an environment of moderate to high productivity, the site will remain open to colonization by seeds or spores for only a relatively short period. The persistence of the primary colonists during the subsequent process of vegetation succession (see Chapter 5) will depend upon the strategy adopted in the established phase. Whereas competitive trees, shrubs, and herbs often persist for a considerable period of time, ruderals are rapidly displaced and depend for their survival upon the occurrence of freshly-disturbed sites within the dispersal range of the population.

Combinations involving a bank of persistent seedlings

The significance of this regenerative strategy in the life-histories of stress-tolerators has been considered already on pages 102–3. It would appear that regeneration involving a bank of persistent seedlings is characteristic of plants adapted to circumstances in which the opportunities for recruitment from the seedling population occur infrequently and depend upon senescence and occasional mortalities among the established plants. Seed production by stress-tolerators is intermittent, a feature which is conducive to long-term survival of the parent plants (page 51). The 'seedling bank' ensures that despite this reduced reproductive effort the capacity for regeneration from seed is maintained between successive seed crops.

Regenerative strategies and breeding systems

In this chapter strategies have been defined in terms of the morphological and physiological characteristics which permit regeneration in different types of environments. Through the concepts of multiple regeneration and regeneration failure an attempt has been made to explore the relationship between ecological amplitude and the number of regenerative strategies functioning in a population or species. It is necessary, however, to add a further dimension to this analysis by examining the inter-relationship between the regenerative strategies and the breeding systems in particular types of life-histories.

Annuals

In the description of the ruderal strategy (Chapter 1) it has been emphasized that because of their short life cycles and high rates of mortality the survival of populations of annual herbs depends upon the ability to produce numerous seeds, and this is associated in many species and populations with the tendency to begin to flower soon after seedling establishment and to maintain seed production even under conditions in which vegetative growth is severely restricted by stress. In many annuals, these characteristics coincide with a high incidence of self-fertilization and it seems reasonable to interpret this as an additional adaptation securing a rapid rate of seed production.

However, although self-fertilization ensures that lack of pollination does not limit the rate of seed production, there is some evidence which suggests that such a breeding system leads to genetic inflexibility which in turn may limit the ecological amplitude of populations and species. Stebbins (1957), for example, has noted that certain annuals of restricted distribution in California, such as *Bromus carinatus* and *Gilia clivorum* maintain 'constant, genetically homozygous pure lines for a large number of generations' and are 'isolated by self-fertilization from other biotypes of the same species.' In contrast Allard (1965) working in the same region with the widely-successful annuals *Avena fatua* and *Bromus mollis* has found that although these annuals are also predominantly self-pollinated their populations are more variable genetically than those of annuals with a restricted ecology. Allard concludes, moreover, that the genetic systems of the successful annuals are more flexible, allowing rapid replication of particular genotypes in uniform and stable environments but retaining the capacity to produce more genetically heterogeneous offspring capable of exploiting more varied habitat conditions 'through selection of types with higher levels of out-crossing, higher crossover rates, or changes in other factors which might increase variability.' Account must be taken also of the fact that many of the most successful annual species accumulate large banks of persistent buried

Plate 1. The rhizomatous storage organs of three species of high competitive ability A, *Epilobium hirsutum*; B, *Pteridium aquilinum*; C, *Petasites hybridus*.

g

Plates 2–5. Vertical sections in early spring (below) and summer (above) through herbaceous vegetation on an alluvial terrace in North Derbyshire, England, dominated by *Epilobium hirsutum* (left) and *Petasites hybridus* (right). In contrast with the stem litter of *E. hirsutum*, that of *P. hybridus* is not persistent and this has allowed colonization by the grass *Poa trivialis* (p) and the annual species *Galium aparine* (g). The inflorescence (f) of *P. hybridus* is a nonphotosynthetic structure which develops in the spring prior to expansion of the leaf canopy.

Plates 6 and 7. Comparison of vegetative and reproductive vigour in two ten-week-old plants of the annual weed *Polygonum persicaria*. Both specimens were grown on the same garden soil but that on

Plates 8 and 9. Two phases in the life-cycle of the large summer annual, *Impatiens glandulifera*, on a river bank in North Derbyshire, England. The plate on the left shows the large seedling emerging in the spring through the stem litter remaining from the previous year whilst that on the right is a vertical section through the canopy of three neighbouring plants photographed in the early summer.

Plate 10. Vertical section in April through the vegetation colonizing a small patch of shallow soil on a limestone outcrop in North Derbyshire, England. The vegetation is composed of fruiting individuals of the small winter-annuals *Arabidopsis thaliana* (A), *Cardamine hirsuta* (C), and Erophila verna (E), together with seedlings of the perennial herbs *Cerastium fontanum* (F), *Dactylis glomerata* (D), and *Plantago lanceolata* (P).

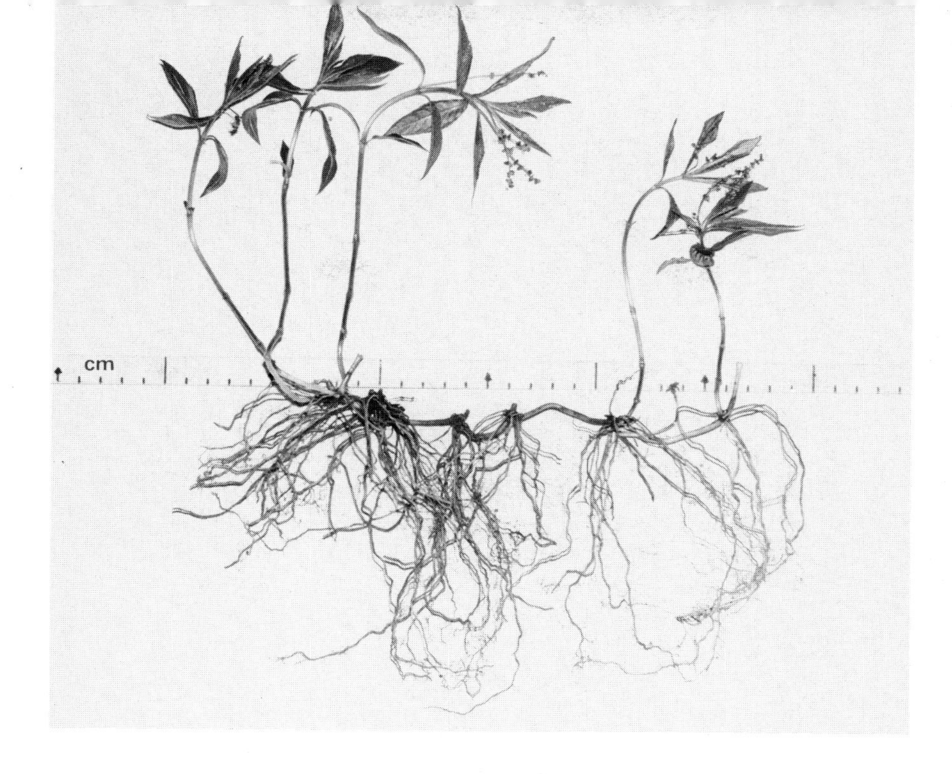

Plates 11 and 12. Photographs illustrating the morphology of two stress-tolerant competitors of contrasted ecology in the British Isles. Above: the calcicolous woodland herb *Mercurialis perennis*. Below: the rhizomatous marshland sedge *Carex acutiformis*.

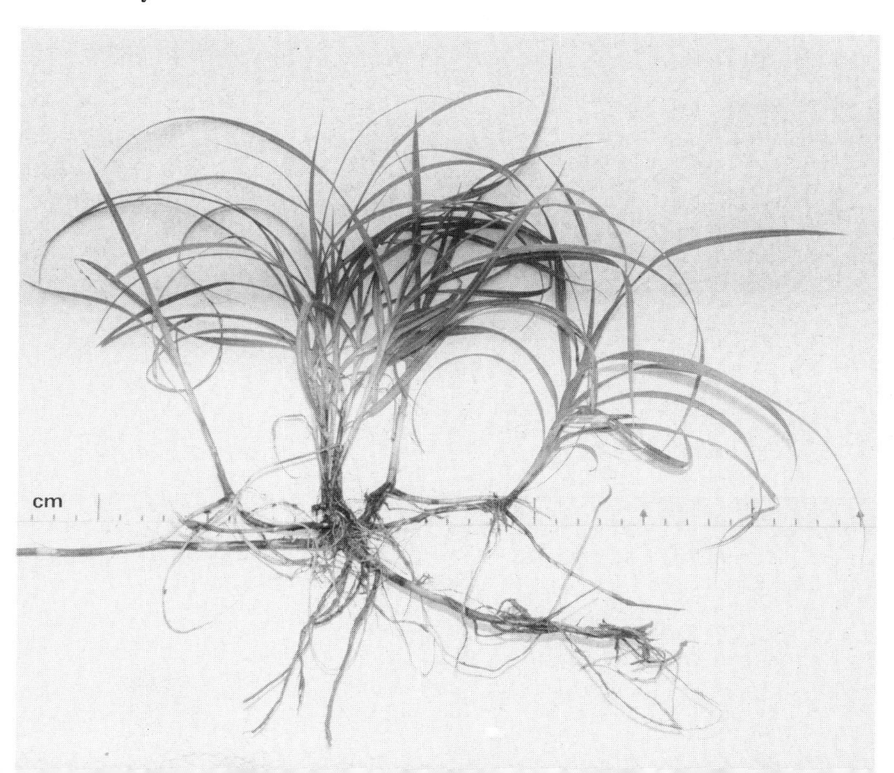

Plate 13. A comparison of two plants of *Poa annua* after five months of growth from seed in the same greenhouse environment. Left: from a derelict limestone quarry. Right: from a cattle pasture. The individual on the left has produced numerous inflorescences (f) and shows widespread senescence and death among the vegetative shoots. No flowering has occurred in the pasture plant which is

'lates 14 and 15. Seedlings which, in Britain, appear in the spring each year in dense, even-aged ›opulations. Above: *Acer pseudoplatanus*. Below: *Endymion non-scriptus*.

Plate 16. The floor of a mixed deciduous woodland in Northern England photographed in early spring to show the interacting effects of topography and persistent tree litter upon the vegetation. On the elevated area near the base of a tree (top left) litter is extremely sparse and cushions of the bryophyte *Mnium hornum* (see Plate 24) occur. Bluebell (*Endymion non-scriptus*) is also present on

Below: *Mnium hornum*, a woodland moss which is ill-adapted for emergence from litter and tends to be restricted to tree stumps (as here), fallen logs, and hummocks.

Plates 25 and 26. Vertically projected view of two plots in a garden transplant experiment designed to investigate the effect of litter upon a mixture of woodland herbs. The photograph illustrates the pattern of shoot emergence at an early stage of the experiment. Left: a control plot containing, in order of decreasing abundance, *Poa trivialis*, *Galeobdolon luteum*, *Holcus mollis*, *Endymion non-scriptus*, and *Viola riviniana*. Right: a plot planted with the same mixture of species and provided *mollissima* and *Quercus rubra*, and two shade-intolerant species, *Ailanthus altissima* and *Rhus glabra*. The seedlings were grown on a continuously moist, fertile loam under full midsummer sunlight over the period June 7–August 29 in New Haven, Connecticut, USA. The grid intervals are 1.27 cm (Grime, 1966). (Reproduced by permission of Blackwell Scientific Publications Ltd.)

Yield (g)	9.6	15.8	6.3	25.3
Seed reserve (g)	2.015	0.009	1.969	0.001
Species	Castanea mollissima	Ailanthus altissima	Quercus rubra	Rhus glabra

Plate 28. Vertical section in April through the vegetation present on a small outcrop of fissured limestone in North Derbyshire, England. Plants of widely different ecology are growing in close proximity, and their co-existence may be related in part to a local gradient (from left to right) of decreasing soil depth. The most conspicuous species are *Dactylis glomerata* (D), *Festuca ovina* (F),

stinging-nettle (*Urtica dioica*) in derelict urban land in Sheffield, England. In spring the copious stem litter of *U. dioica* is almost completely covered by the pleurocarpous moss *Brachythecium rutabulum*.

Plate 31. Dissection of a 25 cm square sample of turf from an ancient limestone pasture in North Derbyshire, England. Left: the intact turf. Right: a sample from the same site sorted into its

Plate 32. The response of four grasses to clipping every two weeks at 5 cm. Left to right: *Lolium perenne* (S23), *Holcus lanatus*, *Agrostis tenuis*, and *Deschampsia flexuosa*. Photograph taken seven days after clipping (Al-Mashhadani and Grime, unpublished).

Plate 33. Two relatively short-lived species which frequently exploit gaps in calcareous grassland arising from drought or biotic disturbance. Left: the potentially large, fast-growing leguminous annual *Medicago lupulina*. Right: the small, slow-growing annual or short-lived perennial *Linum catharticum*.

Plate 34. Comparison of the species density of the turf in two closely adjacent areas of the same sheep pasture in North Derbyshire, England. Above: turf on an acidic (pH 3.5) soil developed on Millstone Grit Sandstone. The species represented are *Festuca ovina*, *Deschampsia flexuosa*, *Agrostis tenuis*, and *Galium saxatile*. Below: turf in an area where the same soil has been subject to calcareous drainage water and the soil is approximately pH 6.5. The main species present are *Festuca ovina*, *Festuca rubra*, *Briza media*, *Agrostis tenuis*, *Trisetum flavescens*, *Carex flacca*, *Trifolium repens*, *Hieracium pilosella*, *Prunella vulgaris*, *Potentilla sterilis*, and *Linum catharticum*.

Plate 35. An epiphytic angiosperm of the genus *Tillandsia* on *Bursera simaruba*, at Royal Palm Hammock, Everglades, Florida, USA.

seeds which will cause an additional source of genetic variability to be injected into the breeding populations each time the habitat is disturbed.

From this example it is evident that genetic considerations are vitally important and must complement the 'functional' insights which arise from the study of regenerative strategies. This point emerges even more clearly in relation to the life-histories of perennial species.

Perennials

Earlier in this chapter it was pointed out that the wide ecological amplitude of perennial species such as *Agrostis tenuis, Holcus lanatus,* and *Trifolium repens* is correlated with possession of several forms of regeneration, and there can be little doubt that the habitat range of these species can be explained in part simply by references to the diversity of conditions under which they are able to regenerate. However, there are additional ways in which multiple forms of regeneration affect the ecological amplitude of a species or population. These arise from the flexibility which is introduced into the genetic system when sexual and asexual forms of reproduction are exhibited by the same genotype, population, or species.

Flexibility is exceptionally high where a predominantly outbreeding method of seed production coincides with an asexual regenerative strategy (e.g. many perennial grasses and forbs and some trees and shrubs). It would appear that in the life-histories of these plants sexual and asexual reproduction play complementary roles. Replication of particular genotypes by asexual propagation allows rapid modification of populations in response to local habitat conditions. Sexual reproduction, on the other hand, dictates a more conservative response to natural selection, since the result of cross-fertilization is (*a*) to introduce among the offspring genes derived from other individuals, many of which may be growing in different types of habitat, and (*b*) to destroy, at meiosis, genotypes of proven selective value in the habitat. It would appear therefore that the effect* of sexual reproduction is to retain genotypic variety within populations, and there can be little doubt that this improves the chance of long-term survival in circumstances where the population remains *in situ* but is exposed to fluctuating conditions (proclimax plants) or depends for its continued existence upon migration to new habitats (successional plants). The contribution of sexual reproduction has been succinctly described by Williams (1975) who states that 'genotypic variety provides a margin of safety against environmental uncertainty,' and argues that 'this may be more important to population survival than is precise adaptation to current conditions.'

We may conclude, therefore, that in perennial species the genotypic variety generated by sexual reproduction is particularly important when regeneration

*Whilst the long-term advantage of sexual reproduction to a population or species is evident, arguments continue with regard to the forces of natural selection which maintain the immediate advantage of sexual reproduction in each generation (see Fisher, 1930; Turner, 1967; Lewontin, 1971; Levin, 1975; Williams, 1975).

or expansion of populations involves establishment in environments which differ substantially from those experienced by the parents. Moreover, sexual reproduction often restores to the species genotypic variety lost in populations undergoing rapid evolution in what Lewontin (1965) has described as 'colonizing episodes'. Colonizing episodes vary in frequency and scale according to the life-history and genetics of the species and the dynamics of the vegetation types in which they occur (Baker and Stebbins, 1965). Nevertheless, all of the species which are capable of colonizing episodes share in common two characteristics. These consist of great fecundity and exceedingly high rates of mortality among seedlings and immature plants. In Britain, for example, many perennial herbs such as *Chamaenerion angustifolium*, *Arrhenatherum elatius*, and *Tussilago farfara*, produce extraordinarily dense populations of seedlings each year and their sudden appearances and disappearances are seasonal events familiar to most field ecologists. This phenomenon also occurs in many perennials which reproduce exclusively by sexual reproduction, including monocarpic herbs such as *Digitalis purpurea*, *Verbascum thapsus*, and trees such as *Betula pubescens*, *Ulmus glabra*, and *Ailanthus altissima*.

On first inspection this excessively wasteful method of reproduction might be interpreted as a consequence of the seed output needed to distribute seeds into the 'safe' sites (Sagar and Harper, 1961) where there is an opportunity for establishment. However, the density of seeds is often far in excess of that required to reach all of the safe sites within the dispersal range and in many instances the effect of over-saturation is to produce dense populations of seedlings within them (see Plates 14 and 15). The most likely explanation for such fecundity is that of Williams (1975) who suggests that it is related to the genotypic variety among sexually-derived progeny and is a measure of the over-production required to ensure a high probability that appropriate genotypes impinge on the safe sites.

Although the most conspicuous examples of high fecundity and seedling mortality in outbreeding species are in plants which produce numerous wind-dispersed seeds, a similar pattern occurs among the perennial herbs and shrubs which form persistent seed banks. The same principles may be used to understand reproduction involving a bank of persistent seedlings although here it is important to bear in mind that the parent plants are often exceedingly long-lived and high values for seed production and mortality become apparent only when they are expressed in relation to the total seed output over the life-span of the parent.

Regenerative strategies in fungi and in animals

There is now an extensive literature concerning the possible adaptive significance of differences in the timing and form of reproductive effort in animal life-histories. Much of this work is theoretical and controversial and, as Stearns (1976) has pointed out, often considerable effort has been devoted to 'the development of ideas for which no one has established connections with the real world.' In this situation, no useful purpose would be served by attempting a detailed comparison of regenerative strategies in plants and animals. It is possible to observe, however, that several of the regenerative strategies recognized in this chapter appear to be closely similar to mechanisms occurring in heterotrophic organisms.

One of the most obvious parallels between plants and animals is that which may be drawn between 'fugitive' herbs and trees and many of the sedentary marine animals (e.g. barnacles, oysters). In both groups migration and establishment in 'safe-sites' is achieved through the production of numerous small, widely-dispersed offspring, most of which suffer early mortality.

Close similarities in regenerative mechanism are also apparent between autotrophic and heterotrophic organisms in environments subject to frequent and temporally unpredictable disturbance. Under such conditions, the role of the seed bank in the survival and re-establishment of plant populations is closely similar to that of the various resting-stages produced by many bacteria, fungi, molluscs, insects, and fish. Here, one of the most remarkable parallels is that between the 'polymorphic' germination requirements of buried seeds (see page 96) and the complex polymorphism which has been described (Wourms, 1972) in the hatching responses of the eggs of annual fish.

In both plants (large-seeded trees and shrubs) and animals (larger mammals and birds) there are regenerative strategies in which risk of juvenile mortality appears to be reduced through the production of a small number of large offspring (Salisbury, 1942; Lack, 1947, 1948; Cody, 1966). It is interesting to note, however, that in functional rather than genetic terms, some of the closest parallels with the large dependent offspring of mammals and birds appear to exist in the asexually-derived progeny of plants regenerating by vegetative expansion.

The complementary roles of sexual and asexual regeneration observed in the life-cycles of many colonizing plants (page 117) recur in certain heterotrophic organisms. They are particularly well known in rust-fungi and in aphids, in both of which colonizing episodes involving the replication of favoured genotypes alternate with sexual forms of regeneration.

PART 2

VEGETATION PROCESSES

Chapter 4

Dominance

INTRODUCTION

For both plants and animals there is abundant evidence of the advantage of large size in many of the interactions which occur either between species or between individuals of the same species (e.g. Black, 1958; Bronson, 1964; Miller, 1968; Grant, 1970; Grime, 1973a). In natural vegetation, examples of the impact of plant size are widespread. A consistent feature of vegetation succession is the incursion of plants of greater stature and at each stage of succession the major components of the plant biomass are usually the species with the largest life-forms. It is necessary, therefore, to acknowledge the existence of another dimension—that of dominance—in the mechanism controlling the composition of plant communities and the differentiation of vegetation types.

In attempting to analyse the nature of dominance it is helpful to recognize two components:

(*a*) the mechanism whereby the dominant plant achieves a size larger than that of its associates; this mechanism will vary according to which strategy is favoured by the habitat conditions;

(*b*) the deleterious effects which large plants may exert upon the fitness of smaller neighbours; these consist principally of the forms of stress which arise from shading, depletion of mineral nutrients and water in the soil, the deposition of leaf litter, and the release of phytotoxic compounds.

It is suggested, first, that dominance depends upon a positive feed-back between (*a*) and (*b*) and, second, that because of the variable nature of (*a*), different types of dominance may be recognized. A broad distinction can be drawn between 'competitive', 'stress-tolerant', and 'ruderal' dominants. In 'competitive dominants' (usually perennial herbs, shrubs, or trees of the early stages of succession in productive habitats) large stature depends upon a high rate of uptake of resources from the environment, whilst in 'stress-tolerant dominants' (usually perennial herbs, shrubs, and trees of unproductive habitats or of the late successional stages of productive habitats) large stature is related to the ability to sustain slow rates of growth under limiting conditions over comparatively long periods. Large stature in 'ruderal dominants' (usually annual herbs) depends upon such characteristics as large seed reserves and rapid rates of germination, growth, and seed production.

The impact of a dominant plant may be exerted upon neighbours at various stages of their life-cycles. Mineral nutrient depletion and shading, together with the physical and chemical effects of litter, are capable not only of preventing seedling establishment but may also reduce the seed output of established plants and bring about premature mortalities among their populations. However, there can be little doubt that the most profound and widespread effects of dominance operate upon the seedling stage. As we shall see later in this chapter, this phenomenon is well illustrated by trees and shrubs, many of which as seedlings are subject to dominance by established perennial herbs but are themselves capable of dominance (often over the self-same herbs) at a later stage of their life-cycles.

DOMINANCE IN HERBACEOUS VEGETATION

Dominance by competitive-ruderals

In severely disturbed productive habitats the vegetation usually consists of a heterogeneous assemblage of ephemeral plants, none of which is capable of functioning as a dominant in the short intervals between successive effects of disturbance. However, where disturbance occurs rather less frequently and conditions favour the persistence of competitive-ruderals (see page 58), dominance may be observed. A competitive-ruderal which habitually attains the status of a dominant is *Impatiens glandulifera* (Plate 9), a large summer annual which in Europe colonizes extensive areas where the margins of water courses have been disturbed by erosion, flooding, and silt deposition. During the summer *I. glandulifera* produces dense colonies and it is not uncommon for other species, including perennial herbs, to be submerged beneath a continuous tall leafy canopy composed of a large number of plants. It is clear therefore that the tendency of *I. glandulifera* to suppress the growth of other species colonizing its habitat is due to the concerted effect of neighbouring plants which, as isolated individuals, would be incapable of exerting a significant degree of dominance.

It is interesting to note that the objective of many forms of arable farming, especially cereal cultivation, is to achieve weed control by creating conditions in which the crop plant attains the status of a dominant. As in the example of *I. glandulifera*, dominance by a cereal crop depends primarily upon the synchronous germination of a high density of large seeds followed by the rapid development of a dense vegetation cover composed of a large number of plants of comparable age and maturity.

It is not unreasonable therefore to describe the events preceding ruderal dominance as a race between seedlings, the outcome of which is measured in terms of relative seed output and is mainly determined firstly by the frequency, size, and germination characteristics of contending seed populations and secondly by the growth-rates and morphologies of the seedlings and established plants.

In the discussion on page 45, reference has been made to the overriding importance of events early in the establishment of ruderal populations in determining the fate of component species. This phenomenon is highly relevant to arable farming where it is desirable to maximize the cost-effectiveness of weeding by selecting the optimal times for mechanical or chemical control measures. Agriculturalists such as Nieto, Brondo, and Gonzalez (1968), Dawson (1970), and Roberts, Bond, and Hewson (1976) have, in fact, recognized a 'critical period', usually extending for a few weeks after crop emergence, during which depletion of resources by the presence of weeds may exercise a major effect upon yield. Provided that weeds are eliminated during the critical period, it has been shown that those appearing subsequently have an insignificant influence upon crop yield. This we may presume to be due to two factors. The first is the tendency of the crop to dominate late-germinating weeds, whilst the second is the 'ruderal' response (i.e. condensation of the life-history) exhibited by weeds subjected to the more limiting conditions experienced during this more advanced stage of vegetation development.

Dominance by competitors

In derelict productive grassland and wasteland where perennial herbs are allowed to grow without major disturbance, there is a well-documented tendency (Watt, 1955; Smith *et al.*, 1971; Grime, 1973a) for certain of the larger plants to expand and to suppress the growth of smaller neighbours. Despite the presence, under such conditions, of a large reservoir of resources, the effect of the activity of the plants during the growing season is to produce expanding zones of depletion, the most conspicuous of which are for light (expanding upwards from the soil surface) and for water and mineral nutrients (expanding downwards from the soil surface). In this type of vegetation high rates of mortality during the growing season and low rates of reproduction are characteristic of those plants which are outstripped by their neighbours and become 'trapped' in the depleted zones. If the growth of the larger species remains unchecked a process of exclusion occurs and this may eventually result in the vegetation approaching a state of monoculture. It seems reasonable to conclude that the dramatic reduction in species densities in meadows and pastures observed over the last thirty years in Europe (e.g. Thurston 1969) is to a large extent the result of an increasing intensity of competitive dominance brought about by stimulating the yield of the more robust and productive species and genotypes through the application of high rates of mineral fertilizers.

The ability of large, fast-growing perennial herbs to suppress the growth of smaller neighbours is particularly evident in vegetation dominated by tall herbs. Dominance by these plants is invariably associated with fertile or moder-

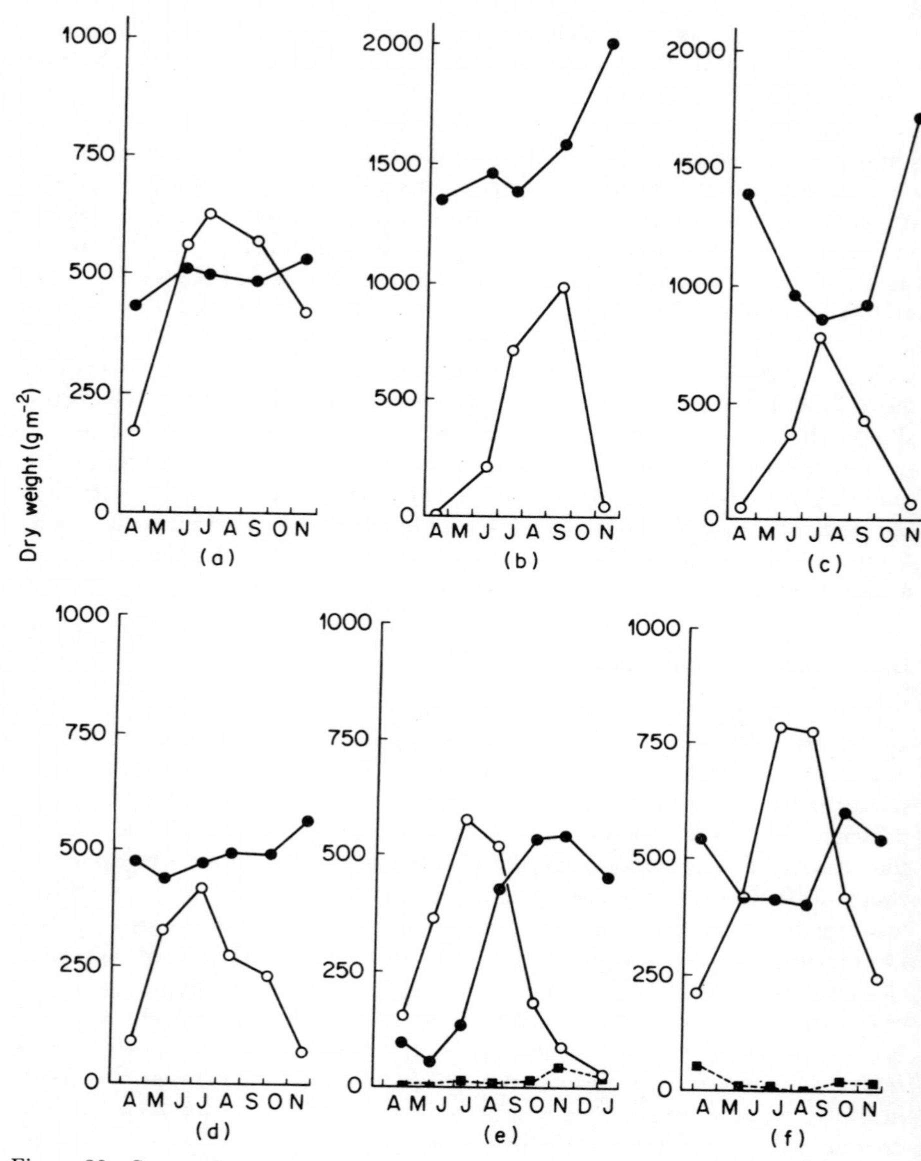

Figure 30 Seasonal change in the total shoot biomass (○), herbaceous litter (●), and tree litter (■) in six tall-herb communities in productive, relatively undisturbed conditions in Northern England. The dominant species at the sites were (a) *Urtica dioica*, (b) *Pteridium aquilinum*, (c) *Chamaenerion angustifolium*, (d) *Filipendula ulmaria*, (e) *Petasites hybridus*, (f) *Urtica dioica*. (Al-Mufti *et al.*, 1977.) (Reproduced by permission of Blackwell Scientific Publications Ltd.)

ately fertile soil conditions and with circumstances in which the vegetation is subject to little disturbance. Studies of the seasonal changes in biomass in tall herb communities (Al-Mufti *et al.*, 1977, Figure 30) reveal that where species such as *Urtica dioica, Pteridium aquilinum, Chamaenerion angustifolium*, and *Petasites hybridus* are dominant there is a high peak (>400 g m^{-2}) in standing crop during the summer and at this time most of the living shoot material is accounted for by the dominant species. As explained in Chapter 1, these plants have in common a number of morphological and phenological characteristics which allow them to maximize the capture of resources during the summer. However, the dominant impact of these species cannot be explained simply in terms of the ability to monopolize the environment during the season of high potential productivity and, in particular, it is necessary to explain why species with shoot phenologies complementary to the dominants (i.e. vernal or evergreen plants) tend to be suppressed or excluded. From data such as those illustrated in Figure 30, there is strong evidence suggesting that a major component of the exclusion mechanism is the presence throughout the year of a high density of herbaceous litter. The impact of litter is particularly well exemplified by the strongly rhizomatous fern *Pteridium aquilinum.* Although it is frost-sensitive and has a very short-lived canopy, this species forms single-species stands over very large areas of derelict land. Invariably, sites dominated by *P. aquilinum* are characterized by the presence of a very dense accumulation (1000–2000 g m^{-2}) of litter which shows comparatively little varation with season and it seems reasonable to conclude that the litter, either by shading or by physical impedence of germination, establishment and growth, restricts the frequency of smaller or slower-growing species. Confirmation of the importance of a constant cover of herbaceous litter in excluding other species may be obtained by reference to another competitive dominant *Petasites hybridus* which, like *P. aquilinum*, produces copious quantities of litter. In comparison with that of *P. aquilinum*, however, the litter of *P. hybridus* is highly palatable to decomposing organisms and, as shown in Figure 30e, declines rapidly to a minimum in the spring, at which time subsidiary species such as *Poa trivialis* and *Galium aparine* often make a temporary but major contribution to the standing crop.

Another observation which suggests the importance of litter in mechanisms of competitive dominance concerns the morphology and vernation of plants such as *Pteridium aquilinum, Urtica dioica*, and *Chamaenerion angustifolium*. Without exception the shoots of these plants are robust structures capable of penetrating a thick layer of litter and expanding the leaf laminae above it. This feature also occurs in a number of tall grasses, e.g. *Arrhenatherum elatius, Alopecurus pratensis*, which are capable of functioning as competitive dominants in productive meadows, derelict grasslands, road verges, and wasteland.

Although the effect of herbaceous litter is apparent at northern and temperate latitudes it is not confined to these regions. A strong impact of litter is described by Egunjobi (1974a) in an area of unburned savannah grassland in Western Nigeria dominated by the tall perennial grasses *Andropogon gayanus* and *Imperata cylindrica.* Measurements at this site (Figure 31) showed that in early spring, at

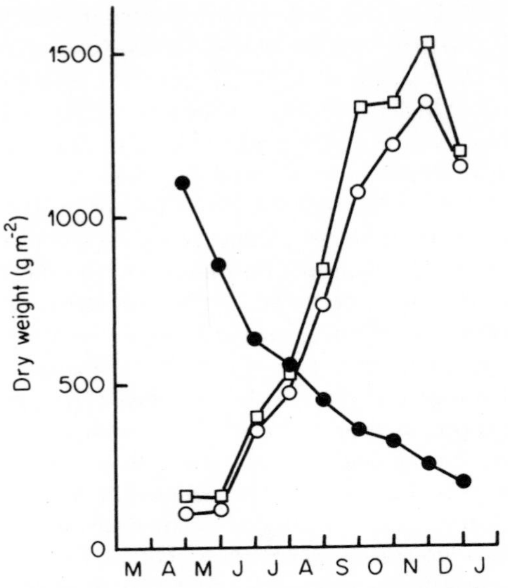

Figure 31 Seasonal changes in the total shoot biomass (□), in the shoot biomass of the tall perennial grass *Andropogon gayanus* (○), and in the dry weight of litter (•) in an unburned savanna grassland in Western Nigeria (Egunjobi, 1974a). (Reproduced by permission of rédaction of *Oecologia Plantarum*.)

the onset of the rainy growing period, the litter density was high (>1000 g m^{-2}) and growth of the vegetation during April and May was retarded 'probably as a result of the choking effect of the litter.'

In many types of herbaceous vegetation competitive dominance is a component of the mechanism controlling species density (number of species per unit area) and in Chapter 6 the practical significance of this fact will be considered.

Dominance by stress-tolerant competitors

Effects of dominance upon species density in herbaceous vegetation are not confined to highly productive environments. Even where grassland or heathland is established on shallow infertile soils, a gradual trend towards monoculture is frequently observed providing that conditions allow the plant biomass to remain relatively undisturbed. Here a well-known example is the widespread fall in species density which has been associated with reductions in grazing intensity by rabbits and sheep in unfertilized calcareous grasslands of the British Isles (e.g. Watt, 1957; Thomas, 1960; Duffey *et al.*, 1974). In this example, the disappearance of the smaller low-growing herbs is correlated with the expansion of grasses such as *Brachypodium pinnatum*, *Festuca rubra*, and *Dactylis glomerata*. As in the case of the dominants associated with more fertile habitats (e.g. *Urtica dioica*, *Epilobium hirsutum*, *Chamaenerion angustifolium*) these species develop a dense leaf canopy over the period June–August and produce a high density of persistent litter.

Dominance by stress-tolerant competitors is particularly common in unfertilized grasslands in semi-arid continental climates. In phenological studies in North Dakota, U.S.A., for example, Redman (1975) has described vegetation in which grasses and sedges such as *Stipa viridula*, *S. comata*, and *Carex pensylvanica* exercise local dominance through moderately-rapid rates of dry matter production and litter accumulation.

An index of dominance

In the preceding sections dominance by various types of herbaceous plants has been associated with particular plant characteristics, e.g. a high leaf canopy and litter accumulation. We may predict, therefore, that in natural vegetation there will be a correlation between the appearance of these characteristics and the tendency for species density to fall as subordinate species are eliminated by dominance. One method of testing this hypothesis relies upon classification of

Table 20. Examples illustrating the derivation of the dominance index.

Species	Attributes				Dominance index (Total/2)
	(a)	(b)	(c)	(d)	
Chamaenerion angustifolium	5	5	5	2	8.5
Arrhenatherum elatius	5	4	4	3	8.0
Brachypodium pinnatum	3	4	3	5	7.5
Ranunculus repens	3	5	3	1	6.0
Helictotrichon pratense	3	2	3	2	5.0
Taraxacum officinale	3	1	4	1	4.5
Festuca ovina	2	1	3	2	4.0
Campanula rotundifolia	2	2	3	0	3.5
Arenaria serpyllifolia	1	0	4	0	2.5

KEY TO SCORING SYSTEM. (a) Maximum plant height: 1, <26 cm; 2, 26–50 cm; 3, 51–75 cm; 4, 76–100 cm; 5, >100 cm. (b) Morphology: 0, small therophytes; 1, robust therophytes; 2, perennials with compact unbranched rhizome or forming small (>10 cm diameter) tussock; 3, perennials with rhizomatous system of tussock, attaining diameter 10–25 cm; 4, perennials attaining diameter 26–100 cm; 5, perennials attaining diameter >100 cm. (c) Relative growth rate (g/g/wk): 1, <0.31; 2, 0.31–0.65; 3, 0.66–1.00; 4, 1.01–1.35; 5, >1.35. (d) Maximum accumulation of persistent (i.e. from one growing season to the next) litter produced by the species: 0, none; 1, thin, discontinuous cover; 2, thin, continuous cover; 3, up to 1 cm depth; 4, up to 5 cm depth; 5, >5 cm depth.

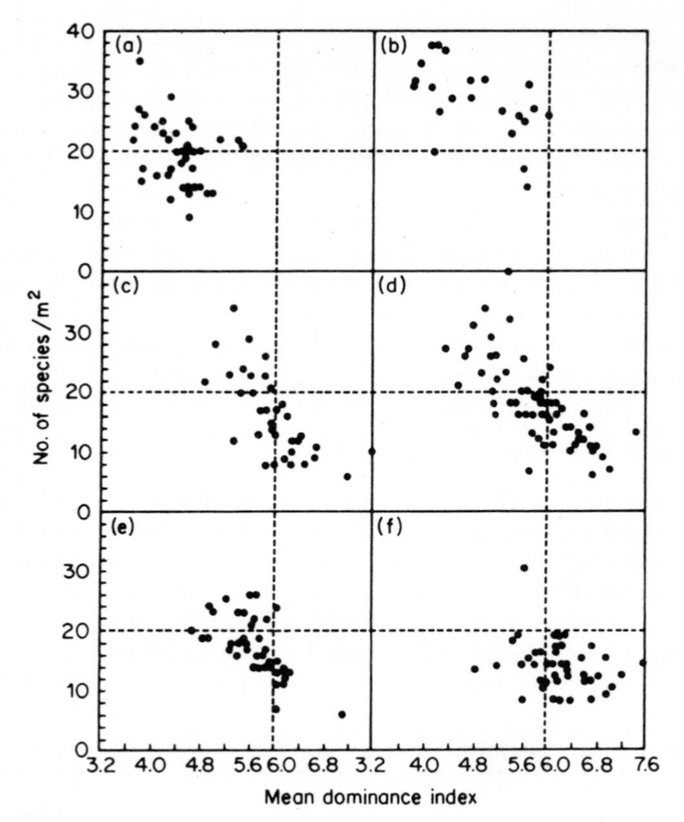

Figure 32 Relationship between a dominance index and species density in herbaceous vegetation in six habitats sampled widely from the same geographical area in Northern England. (a) Limestone outcrops with discontinuous soil cover, (b) unenclosed limestone pastures; (c) enclosed pastures; (d) derelict limestone grassland; (e) meadows; (f) road verges. The contribution of each species to the mean dominance index is weighted in proportion to its frequency in the m² sample of vegetation (Grime, 1973a). (Reproduced by permission of Macmillan (Journals) Ltd.)

herbaceous plants by means of a dominance index* (Grime, 1973a). The dominance index illustrated in Table 20 is based upon four plant attributes consisting of the observed maxima for the species in height of leaf canopy, lateral spread, relative rate of dry matter production (under standardized productive laboratory conditions), and extent of litter accumulation. Each species has been scored with respect to the four attributes and the sum of scores has been used to derive a dominance index with a scale from 0–10.

*This index was originally restricted in application to vegetation experiencing competitive dominance and was for this reason described as a competitive index.

Although it ignores genetic variation within the species with respect to the potential for dominance and is rather subjective, especially with regard to the estimation of the potential for litter accumulation, the dominance index appears to be informative. In Figure 32 the dominance index has been used to seek evidence of effects of dominance upon species density in several types of vegetation sampled from the same geographical area in the north of England. For each square metre of herbaceous vegetation examined, a mean value has been calculated from the dominance indices of the component species and the contribution of each species to the mean has been weighted according to the frequency of the species in the vegetation sample. In Figure 32, the mean dominance index for each sample has been plotted against the species density (number of species per square metre). The results show that where species of high dominance index are abundant (that is, where the mean D.I. exceeds 6.0) species densities are relatively low (less than 20 species per m^2). From the data presented in Figure 32a,b it would appear that in the area of study there is a low incidence of exclusion of species by dominance in the sparse vegetation characteristic of sheep pastures and rock outcrops over limestone. In contrast, there is strong evidence that dominance is a causal factor in the maintenance of the rather low species densities encountered in the samples from road verges (Figure 32f). The three remaining vegetation types examined in Figure 32 are enclosed pastures, derelict limestone grassland, and meadows. In each, species density varies widely but shows a marked decline with increasing dominance index.

These results appear to confirm that the development of a marked degree of dominance in herbaceous vegetation is associated with the appearance of a high frequency of particular plant characteristics. For the practical purposes of vegetation management it seems possible, therefore, that criteria such as those used in the dominance index may be used to assess or anticipate the impact of dominance either at individual sites or in different types of vegetation.

DOMINANT EFFECTS OF HERBACEOUS PERENNIALS UPON TREE SEEDLINGS

With the exception of environments subjected to continuous or repeated disturbance there is a tendency for herbaceous vegetation to be colonized by the seedlings of shrubs and trees. The most obvious advantage of seedlings of woody species establishing in vegetation dominated by perennial herbs is the cumulative nature of height growth which, in successive seasons, elevates the leaf canopy of the tree seedling above that of the tallest herbs. However, despite their rapid growth in height, tree seedlings and saplings may become severely stunted in the presence of herbaceous species such as *Pteridium aquilinum* and *Chamaenerion angustifolium* which, because of their rapid growth-rates, capacity for lateral spread above and below ground, and copious litter production, may subject tree seedlings to major effects of dominance. The ability of perennial herbs to retard the growth of tree seedlings and saplings is well known to foresters and it is now common practice, as a preliminary to tree planting, to check the growth of adjacent herbs by treatment with herbicides.

Apart from their practical and economic importance, the strong effects which productive herbaceous vegetation may exert upon the growth of tree seedlings are of considerable theoretical interest, particularly in relation to the process of vegetation succession (see page 148). Three characteristics may be identified, which may account for the vulnerability of many tree seedlings. These are: low relative growth rate, delay in consolidation and lateral spread of the leaf canopy, and failure at least during the initial phase of establishment to accumulate a deep and persistent layer of leaf litter. Of these characteristics, the most consistent is the low relative growth rate which usually falls well below that of the seedlings of perennial herbs (Jarvis and Jarvis, 1964; Grime and Hunt, 1975; Ampofo *et al.*, 1976). The comparatively slow growth of tree seedlings has been attributed to the expenditure of photosynthate on woody tissue, a process concomitant with a slow rate of expansion of leaf area (Jarvis and Jarvis, 1964).

As explained on page 139, a characteristic which appears to increase the chance of successful tree seedling establishment in vegetation composed of perennial herbs is the possession of a large seed.

DOMINANT EFFECTS OF TREES UPON HERBACEOUS PLANTS

Some of the most dramatic effects of trees upon herbaceous vegetation occur in commercial forests and plantations. In mature, even-aged stands, both coniferous and deciduous species are capable of developing an extremely dense canopy and it is not unusual in these circumstances for the regeneration of herbaceous and woody species to be suppressed and for the herb layer to be extremely sparse. A particularly good system in which to detect the impact of trees upon herbaceous vegetation occurs in coppiced woodlands where the tree canopy is removed at regular intervals varying approximately between 10 and 20 years, and there are associated cycles of regeneration and suppression in the ground flora. Sampling at various stages in the coppice cycle (Figure 33) reveals a close correlation between the biomass of coppice shoots and the level of production in the herb layer.

In order to examine the impact of trees upon herbaceous plants in more natural types of vegetation it is instructive to measure the seasonal patterns of shoot production in the herb layer within a wood and to compare the results with those obtained from concurrent measurements on herbaceous vegetation situated in an adjacent clearing. Such a comparison is illustrated in Figure 34, which refers to a quantitative study of shoot phenology at two sites within a mixed deciduous woodland on an alluvial terrace in Northern England.

The data in Figure 34a show that in the absence of trees the herbaceous vegetation at this site is dominated by one species, *Urtica dioica*, which develops a dense canopy of living shoots during the period of maximum potential productivity (June–August) and produces a large quantity of persistent litter. Beneath

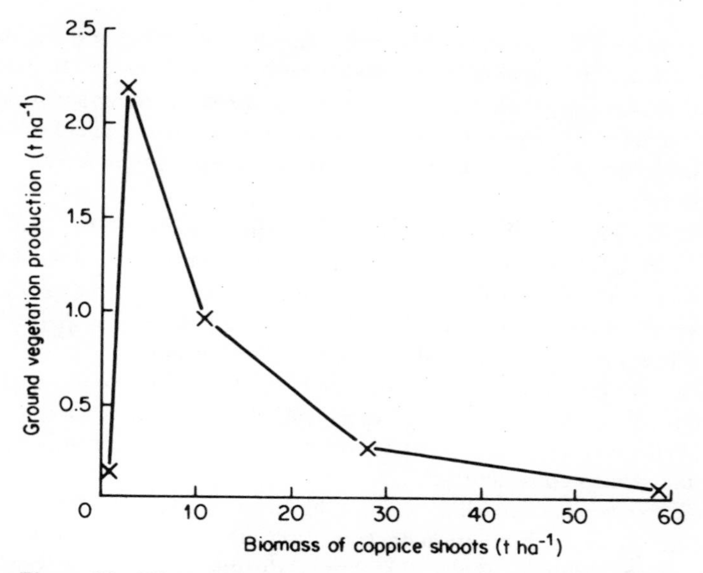

Figure 33 The relationship between annual production in the herb layer and biomass of the coppice shoots at different stages of the coppice cycle in a sweet chestnut (*Castanea sativa*) stand in South-eastern England (Ford and Newbould, 1977). (Reproduced by permission of Blackwell Scientific Publications Ltd.)

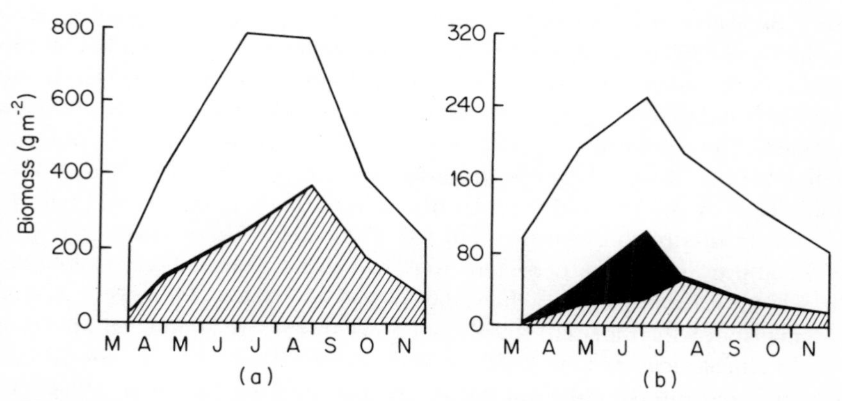

Figure 34 Comparison of the seasonal development of the shoot biomass of *Urtica dioica* (▨) at two adjacent sites on an alluvial terrace in Northern England: (a) in a woodland clearing with no overhead shade, (b) under a deciduous canopy of small trees and shrubs. Also included in both figures is the shoot biomass of the vernal species *Poa trivialis* (■) and that of the remaining herbaceous species (□). (Al-Mufti *et al*., 1977.)

the trees (Figure 34b), however, the standing crop of herbs, and the density of herbaceous litter, is considerably reduced and the peak in dry matter occurs before full expansion of the tree canopy. *Urtica dioica*, although still present, is severely restricted in vigour and is associated with a variety of other woodland herbs. The major contributor to the herbaceous vegetation is *Poa trivialis*, a vernal species.

It would appear, therefore, that, in the example considered, the main effect of the trees is to debilitate the potentially dominant component of the herb layer (*Urtica dioica*) with the result that there is an incursion by a variety of herbaceous species which, although of smaller size and lower competitive ability than *U. dioica*, are better adapted to woodland conditions.

The main reason for the suppression of *Urtica dioica* in woodland appears to be the coincidence between its leaf expansion and that of the trees, although additional effects due to the activity of tree roots and to the presence of tree litter must be taken into consideration.

In some temperate woodlands (e.g. Figure 13, page 62) there is a very pronounced seasonal fluctuation in tree litter, the density of which remains relatively high (approximately 200 g m^{-2}) during the winter, only to fall extremely sharply in the early spring to a minimum of about 10 g m^{-2}. It is not uncommon, however, for the density of tree litter in temperate woodlands to attain seasonal maxima considerably greater than this and, in both deciduous and coniferous woods, there are local situations in which the density of tree litter exceeds 500 g m^{-2} throughout the year. From both field studies and experiments there is evidence that high densities of tree litter exert a profound effect upon the structure and species composition of woodland herb layers.

In a recent investigation (Sydes and Grime, 1979), studies have been made of the response of various woodland herbs to differences in the quantity and type of litter deposited upon them by deciduous trees. Experimental data such as those in Figure 35 reveal that there are marked differences between herbaceous plants in their capacity to tolerate the physical impact of tree litter. In the five species investigated, the order of tolerance was *Galeobdolon luteum*, *Endymion non-scriptus*, *Viola riviniana* > *Holcus mollis* > *Poa trivialis*, and this sequence matches closely the correlations recorded in natural woodland between the abundance of the species and the quantity of tree litter (Figure 36 and Plate 16). From the results of this investigation and from field observations, it is clear that the ability to tolerate persistent tree litter is related to shoot morphology (Plates 17–24). Emergence is achieved in *Galeobdolon luteum* (Plate 22) by creeping stems from which a small number of erect shoots are supported above the litter. In *Allium ursinum* and *Endymion non-scriptus* (Plates 19, 20) the litter is penetrated by robust enciform leaves whilst in *Viola riviniana* (Plate 18) emergence depends upon the combined effect of an erect shoot-base and elongated petioles. The susceptibility of species such as *Poa trivialis* and *Holcus mollis* to persistent tree litter appears to be due to the fact that both the shoots and the individual leaves are insufficiently robust to resist physical impedance and to escape from the shading effect of dense litter.

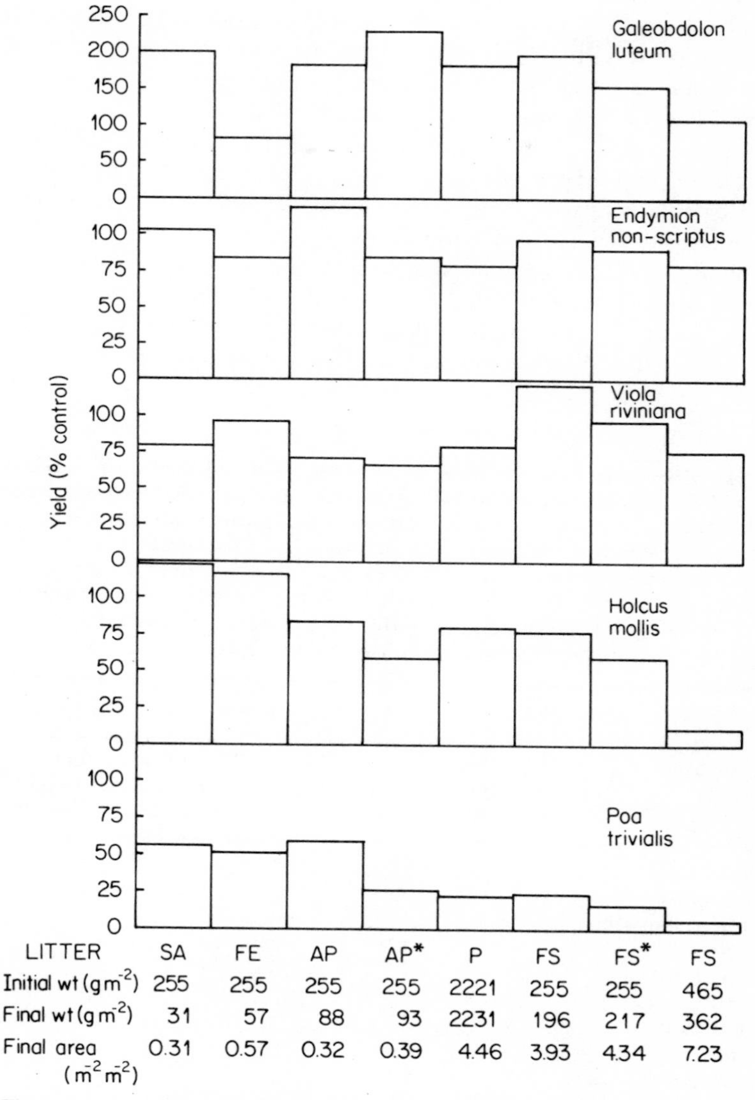

Figure 35 Comparison of the effect of applications of various types and quantities of litter upon the yield of transplants of five herbaceous plants of common occurrence in woodland. Key to litter types: SA, *Sorbus aucuparia* (fresh); FE, *Fraxinus excelsior* (fresh); AP, *Acer pseudoplatanus* (fresh); AP*, *A. pseudoplatanus* (decayed); P, black plastic 'litter' (see Plate 26); FS, *Fagus sylvatica* (fresh); FS*, *F. sylvatica* (decayed). (Reproduced by permission of Blackwell Scientific Publications Ltd.)

It seems likely that physical effects of tree litter upon forest herbs will be restricted to northern and temperate regions. In a number of studies (Jenny *et al.*, 1949; Olsen, 1963; Hopkins, 1966; Egunjobi, 1974b) very rapid rates of litter decomposition in tropical forest have been measured and in certain sites litter has been found to disappear completely within periods varying from 2–7 months (Nye, 1961; Madge, 1965; Bernhard, 1970).

So far discussion of the effects of trees upon herbaceous plants has been restricted to those arising from depletion of resources or the physical impact of litter. However, in a wide range of studies which have been reviewed by Whittaker and Feeny (1971), Muller and Chou (1972), and Rice (1974) evidence has been obtained of circumstances in which compounds released from living foliage or litter have been found to be capable of inhibiting seed germination and/or plant growth. Research workers differ widely in the extent to which they have attributed a direct ecological significance to these phenomena. Major

Table 21. Comparison of the litter of eighteen trees and shrubs of common occurrence in Britain with respect to (a) weight loss during initial period of decay and (b) inhibitory effect of litter leachates on plant growth. Measurements of decay and the preparation of leachates both involved litter maintained in a moist aerobic condition at warm temperatures (20°C day, 15°C night) over a three week period immediately following leaf fall. Growth inhibition was based upon reductions in yield of seedlings of the perennial herb *Rumex acetosa*, grown for 3 weeks in nutrient-sufficient sand culture (Sydes and Grime 1978).

Source of litter	(a) Weight loss by decay and leaching (% dry weight)	(b) Phytotoxic effect of leachate (% reduction in yield of *Rumex acetosa* seedlings)
Sambucus nigra	57	50
Frangula alnus	52	45
Thelycrania sanguineus	46	45
Ulmus glabra	42	31
Fraxinus excelsior	42	31
Alnus glutinosa	33	19
Acer campestre	31	42
Salix caprea	30	10
Acer pseudoplatanus	28	39
Corylus avellana	25	3
Populus canescens	24	17
Betula pendula	23	42
Tilia cordata	23	7
Quercus robur	23	7
Sorbus aucuparia	23	44
Castanea sativa	17	31
Crataegus monogyna	17	18
Fagus sylvatica	15	16

difficulties arise when an attempt is made to determine whether such effects arise from natural selection for allelopathic ability or are merely a consequence of quite different adaptations such as the production of compounds which provide a defence against predation or microbial attack.

An example of the difficulties in the interpretation of quasi-allelopathic effects can be seen from Table 21, which presents the results of an experiment in which litter from a range of common British trees and shrubs was compared with respect to the rate of decay and the ability of leachates to inhibit seedling growth. From these data it is apparent that, in short-term experiments with seedlings, phytotoxic effects can be detected in leachates from litter of species such as *Fraxinus excelsior*, *Sambucus nigra*, and *Sorbus aucuparia*. However, release of toxins from these species appears to be a transient effect associated with the rapid decay and high initial efflux of solutes from freshly-deposited litter. It is clear that the species (e.g. *Fagus sylvatica*, *Quercus robur*) which form the main residue of persistent litter in deciduous woodlands are relatively non-toxic. Moreover, even in *Acer pseudoplatanus*, which combines moderate toxicity with an intermediate degree of persistence, it would appear that the main impact of the litter is due to physical rather than chemical causes. This conclusion is prompted by the experimental evidence in Figure 35 which suggests that freshly-collected litter of *A. pseudoplatanus* was less pronounced, in its effect on herbaceous plants, than litter which, prior to use in the experiment, had experienced leaching and decomposition in a woodland floor for a period of five months. Further evidence suggesting the overriding importance of the physical impact of tree litter is the fact that the deleterious effects of persistent types of tree litter in woodland herbs are closely matched (Figure 35, Plates 25 and 26) by those produced by an equivalent area of artificial litter composed of opaque plastic 'leaves'.

DOMINANT EFFECTS OF TREES UPON OTHER TREES AND SHRUBS

Most of the evidence and theories concerning the interactions between woody plants are based upon field observations in woodlands and plantations (Watt, 1919; Wilde and White, 1939; Ashton and Macauley, 1972; Vaartaja, 1952; Vaartaja and Cran, 1956; Wardle, 1959). These studies have provided insights into the mechanisms whereby the presence of established trees affects the regeneration of trees and shrubs and they have focused attention on two critical phenomena. The first, known to foresters as the 'seed-bed condition', refers to the physical nature of the substratum in which the seed has germinated. The second concerns the role of fungi in seedling mortality.

In view of the profound effects upon woodland herbs described under the last heading it is not surprising to find that tree litter exerts a major influence upon the establishment of the seedlings of trees and shrubs. In temperate woodlands, seedling establishment in relatively small-seeded genera, e.g. *Betula*,

Pinus, Salix, and *Tsuga*, is adversely affected by a continuous cover of tree litter and there is a strong tendency for seedling survival in such species to be restricted to local areas where mineral soil remains exposed or to emergent sites such as rotting logs or tree stumps (Keever, 1973). A marked contrast to this phenomenon occurs in the case of larger-seeded species such as *Quercus, Castanea, Carya*, and *Fagus*, many of which are capable of emergence from beneath litter and appear to experience lower rates of seed predation in such circumstances. This characteristic applies even to certain species with more modest seed reserves; an example of this phenomenon is provided in Figure 36 which illustrates the direct correlation between the density of tree litter and the frequency of seedlings of *Fraxinus excelsior* on the floor of a mixed deciduous woodland in Northern England.

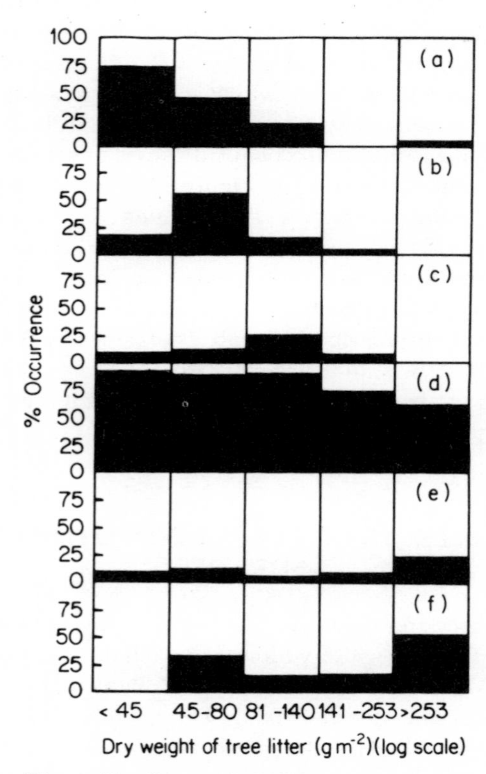

Figure 36 Comparison of the relationship between the shoot biomass of various ground flora components and the density of tree litter present in randomly located samples in deciduous woodland in Northern England: (a) bryophytes, (b) *Holcus mollis*, (c) *Poa trivialis*, (d) *Endymion non-scriptus*, (e) *Galeobdolon luteum*, (f) seedlings of *Fraxinus excelsior* (Sydes, 1979).

From studies such as those of Vaartaja (1952) and Wardle (1959) it is evident that fungal pathogens account for a high proportion of the fatalities which occur among seedlings germinating in temperate woodlands and there is evidence that small seedlings may be predisposed to fungal attack by the low light intensities and high humidities which arise in deep litter and beneath dense leaf canopies of tree and herb layers. 'Damping off' is especially prevalent among the etiolated seedlings of small-seeded, potentially fast-growing trees such as *Betula populifolia* and *Betula lenta* (Grime and Jeffrey, 1965). The capacities to emerge through shaded strata, such as those created by tree litter or herbaceous canopies, and to resist fungal attack are considerably greater in the large-seeded species which occur as the dominants of mature forest (Figures 37, 38).

In addition to effects of shading and fungal attack a variety of other possible mechanisms have been considered (e.g. Janzen, 1970; Fox, 1977) whereby trees might exercise a controlling influence on the regeneration of woody species. The phenomena suggested here include alterations in the availability of soil moisture and mineral nutrients, release of organic toxins from leaf canopies, and encouragement of insect populations which destroy seeds and seedlings. Only when the results of further research are available will it be possible to judge the validity of these hypotheses.

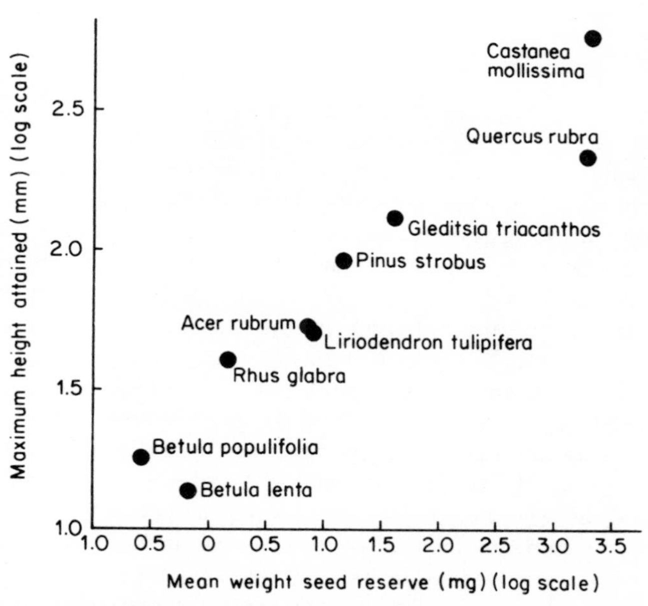

Figure 37 The relationship between weight of seed reserve and maximum height attained after 12 weeks in dense shade by seedlings of nine North American tree species (Grime and Jeffrey, 1965). (Reproduced by permission of Blackwell Scientific Publications Ltd.)

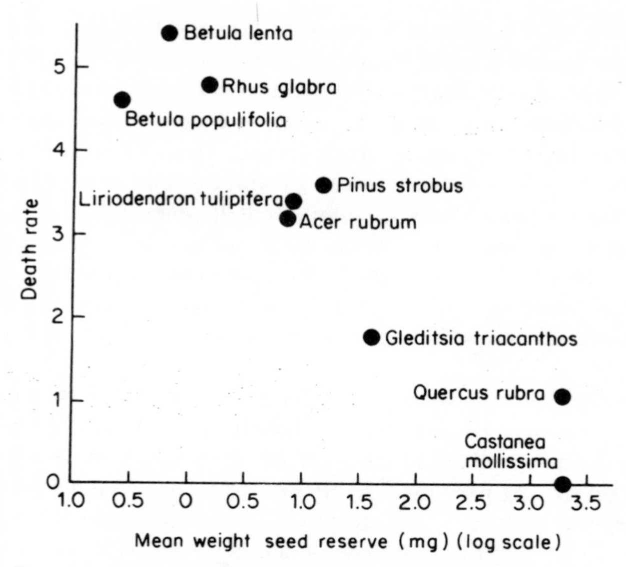

Figure 38 The relationship between weight of seed reserve and death-rate (mean number of fatalities per container in 12 weeks) in dense shade in seedlings of nine North American tree species (Grime and Jeffrey, 1965). (Reproduced by permission of Blackwell Scientific Publications Ltd.)

Seedling mortalities represent only one stage of the process by which the establishment of shrubs and trees may be affected by the presence of mature trees. As explained in Chapter 3, regeneration in many species depends upon exploitation of gaps in the canopy arising from windfalls and senescence. During the colonization of gaps there may be intense competition between saplings, some of which may have originated from banks of persistent seedlings, others from wind-dispersed seeds. In some forests, interactions between the species may result in two or more phases in the recolonization of the gap. In oak–birch woodlands of Northern Britain, for example, gaps in the canopy are usually rapidly colonized by the wind-dispersed species, *Betula pubescens*, the saplings of which grow relatively rapidly and enjoy a temporary advantage over the slower-growing saplings of oak (*Quercus petraea*). In the long term, however, oaks usually replace the birches, and shading appears to be involved in this process. The ability of the oaks to suppress saplings of neighbouring birch trees appears to be related not merely to the fact that it is a taller, longer-lived species. It would seem also that the greater lateral spread and density of the leaf canopy causes the species to cast a deeper, more extensive shade than that of birch, which has a narrow and rather diffuse canopy.

A similar pattern of species replacement occurs during the re-vegetation of gaps in northern hardwood forests of the United States. In the sequence illustrated in Figure 39, for example, the primary colonist, pin cherry (*Prunus pensyl-*

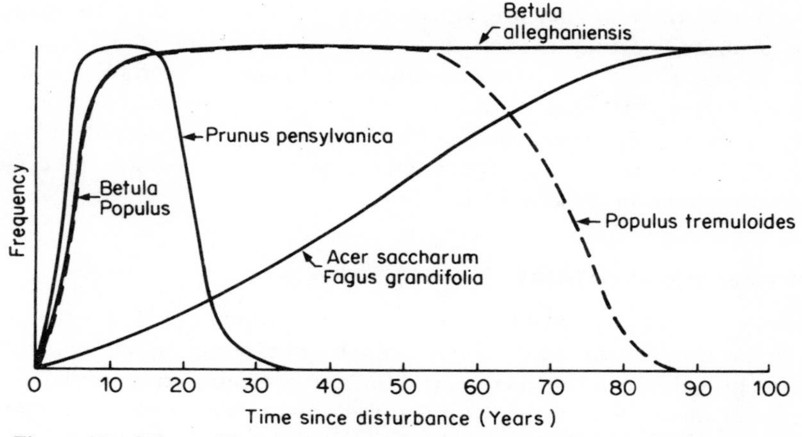

Figure 39 Scheme illustrating a common sequence of vegetation recolonization following local disturbance in northern deciduous woodland in North America (Marks, 1974). (Reproduced by permission of the Ecological Society of America. Copyright 1974 by the Ecological Society of America.)

vanica), is a short-lived species which is excluded in the later stages of recolonization by the arrival of larger, long-lived forest trees.

Shading of pioneer trees and shrubs by forest dominants is also a conspicuous feature of the later stages in the process of woodland development on abandoned fields in the Eastern United States. In this habitat primary colonists such as red cedar (*Juniperus virginiana*) and grey birch (*Betula populifolia*) display extreme sensitivity to shade (Lutz, 1928). As the forest cover closes, shaded sectors of the canopies show rapid die-back and this is followed by high rates of mortality among established trees.

CONCLUSIONS

Success and failure in dominant plants

At the beginning of this chapter dominance was defined by reference to the stresses which are produced by large plants and the effects of these stresses upon smaller neighbours. It is clear, however, that dominance depends not only upon the generation of stresses but also upon the capacity to avoid or resist their effects. Here, a familiar example is the ability of the shoots of herbaceous dominants such as *Pteridium aquilinum* and *Chamaenerion angustifolium* to penetrate through the litter which they have themselves accumulated.

In the 'competitive' dominants of productive environments, stresses originating from local depletion of resources are avoided by means of growth responses which permit a redistribution of the absorptive surfaces of the plant and exploitation of new reserves within the habitat (see page 13). However, as the dominant plant enlarges in size, and resources (especially mineral nutrients) become more scarce as they are accumulated in living and dead components of the biomass, we may predict that the competitive dominants will often decline in vigour and will tend to be replaced by 'stress-tolerant' dominants which are better adapted to withstand such 'autogenic' stress. Phases of dominance, decline, and replacement are, in fact, frequently observed in nature (Watt, 1947) and are an integral part of the processes of succession and cyclical change considered in the next chapter.

Dominance and allelopathy

Both living and dead plants have constituents which either directly or by microbial transformation in the soil are capable of exerting an inhibitory effect on plant growth. Many of these compounds, e.g. lignin, are essential to the structure or metabolism of the plant and their phytotoxic properties appear to be incidental.

In addition there are 'secondary' chemicals in plants, some of which when released into the environment inhibit germination or plant growth. Since many of these substances have no known function within a plant it is tempting to conclude that their primary function is allelopathic. However, as we have seen already (pages 37–39), potentially phytotoxic compounds are often produced as a defence against animal predators or pathogens. The need for caution in attributing an allelopathic role to these substances has been further emphasized by Siegler and Price (1976) who point out that secondary compounds such as terpenes (Loomis, 1967) and alkaloids (Robinson, 1974) often exist in a state of dynamic equilibrium within the plant and are not merely static end-products of metabolism. These authors conclude that many potentially-toxic secondary chemicals in plants have an as yet unknown metabolic function and have not necessarily evolved either to repel herbivores or to inhibit pathogens or competitors.

A further difficulty which must be resolved before allelopathy can find a secure place in ecological theory is that of autotoxicity. From field investigations (e.g. Milton, 1943; Curtis and Cottam, 1950; Voight, 1959; Webb, Tracey, and Haydock, 1967; Hanes, 1971; Tinnin and Muller, 1972; Newman and Rovira, 1975; Al-Mashhadani, 1979) and experimental evidence such as that in Figure 40 it is apparent that many phytotoxic compounds are effective against the plants which produce them. This does not mean, however, that we should assume that these plants derive no advantage from the production and release of toxins. Quite clearly, contingencies may arise in nature which differ substantially from those occurring in laboratory experiments or at local field sites.

One set of conditions in which a plant may escape the harmful effect of its toxins is that in which there is spatial separation between the root system and the zone of contaminated soil. An example of this phenomenon occurs in the chaparral of North America where toxins originating from the foliage and leaf litter of shrubs such as *Adenostoma fasciculatum* tend to be confined to the surface

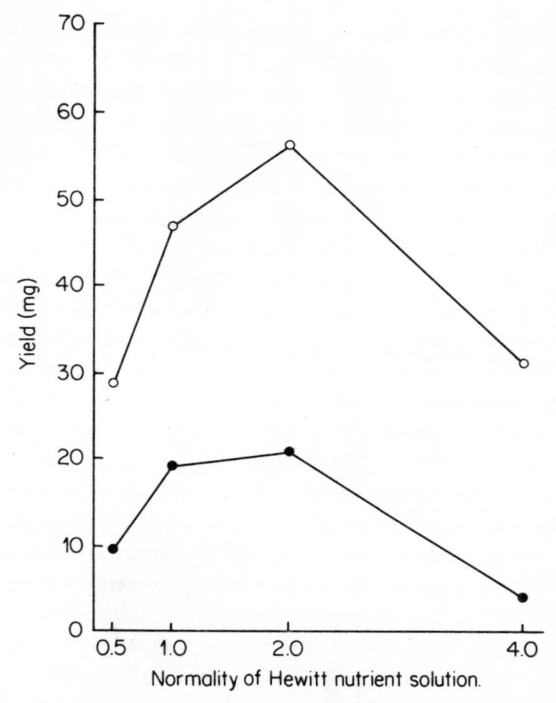

Figure 40 Autotoxicity in *Holcus lanatus* indicated as the yield of seedlings of *Holcus lanatus* after six weeks' growth in sand cultures, supplied with leachates from pots of sand containing no plants (○) and established plants of *H. lanatus* (●). In order to ascertain that the effect of *H. lanatus* was not due to depletion of the level of mineral nutrients supplied in the leachate, the experiment was conducted at four rates of nutrient supply. The nutrient concentrations attributed to the four treatments refer both to the normality of the Hewitt nutrient solution which was supplied directly to the seedlings throughout the experiment and to the solution which was used to elute the leachates (Al-Mashhadani, 1979).

layers of the soil profile. However, there is evidence (McPherson and Muller, 1969; Hanes, 1971) that, in mature chaparral, the level of toxin accumulation reaches a point where autotoxicity occurs and regeneration of *A. fasciculatum* is inhibited. In the normal cycle of events in chaparral, the toxins are destroyed by fire and after an initial phase of colonization by annual plants and grasses there is a phase of renewed toxin accumulation and dominance by shrubs. It is interesting to note that cycles of colonization, dominance, and reduced productivity have been recognized in acidic heathlands and grasslands of the British Isles (Watt, 1947, 1955; Grime, 1963; Grubb, Green, and Merrifield, 1969) although, in these vegetation types, it remains to be shown what part, if any, is played by phytotoxins.

Another circumstance in which toxins may facilitate dominance with minimal risk of autotoxicity is that in which the growth of the toxin-producer is restricted to one season in the year during which toxins accumulate either by secretion or as decomposition products from senescent roots and shoots. Under these hypothetical conditions (Figure 41) the highest concentrations of toxin in the environment will occur after the main growth-period of the producer and may be expected therefore to exert a selective effect against species with complementary phenologies. However, a continued advantage can be expected from such a mechanism only provided that the rate of breakdown of the toxin is sufficient to reduce the concentration to a low level before the next growth-period of the toxin-producer and, as shown in Figure 41, alteration of the balance between rates of production and decomposition of toxin could result in conditions favourable either to co-existence or to autotoxicity. At the present time, there is only circumstantial evidence (e.g. Robinson, 1971) of such 'seasonal' allelopathy. However, it may be significant that phytotoxic effects have been reported for a considerable number of competitive-ruderals, many of which achieve local or temporary dominance in herbaceous vegetation. Examples include annual forbs, such as *Hordeum vulgare* (Pickering, 1919; Overland, 1966), *Camelina alyssum* (Grümmer and Beyer, 1960), *Helianthus annuus* (Wilson and Rice, 1968), and *Avena fatua* (Tinnin and Muller, 1972), grasses such as *Lolium multiflorum* (Welbank, 1963) and *Holcus lanatus* (Al-Mashhadani, 1978), and perennial forbs such as *Trifolium* spp. (Chang *et al.*, 1969; Katznelson, 1972) and *Andropogon virginicus* (Rice, 1972).

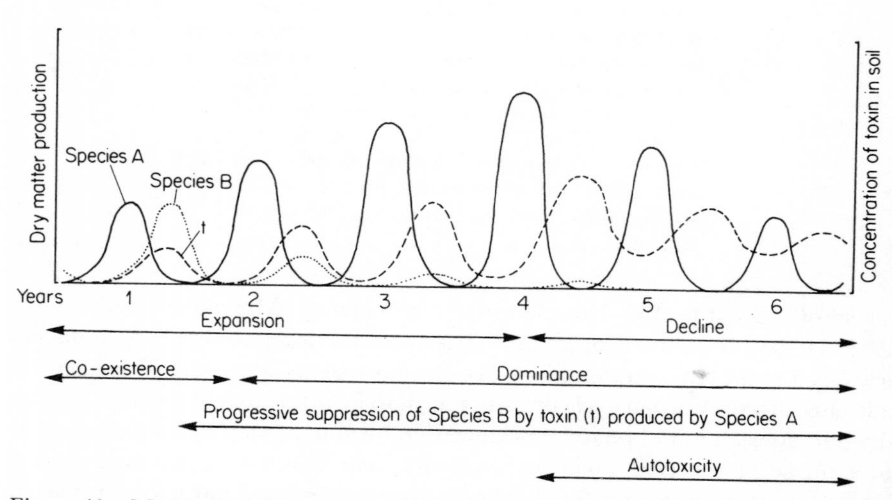

Figure 41 Model describing the possible role of accumulating plant toxins in mechanisms of co-existence, dominance, and autoxicity involving two herbaceous species.

Co-existence with dominant plants

It is unusual for the process of dominance to result in stands of vegetation composed of a single species. Even in circumstances where most of the biomass belongs to one species it is not uncommon for others to remain as a relatively minor but constant component of the vegetation. Studies of these plants suggest that there is a variety of mechanisms whereby subordinate species may co-exist with the dominants. Some, by virtue of their morphology or phenology, escape the main impact of the stresses generated by the dominant (e.g. vernal herbs beneath deciduous trees) whilst others (e.g. evergreen herbs in woodland) are adapted to tolerate these stresses. Co-existence in certain other plants is achieved by exploitation of areas within the habitat where the environment is locally unfavourable to the dominant. In addition there are subordinate species which owe their presence to micro-habitats created by the dominant (e.g. epiphytic angiosperms, ferns, mosses, and lichens). A more detailed examination of these phenomena is attempted in Chapter 6.

Analogous phenomena involving animals

Dominance has been the subject of many investigations by animal ecologists, and the results of these studies suggest that different mechanisms of dominance, resembling in certain respects those observed in plants, can be recognized.

'Ruderal' dominance is apparent in populations of certain birds, rodents, and insects in which relatively small animals acting in concert exercise a dominant, if temporary, effect upon the fate of other animals exploiting the habitat. Familiar examples here include insect eruptions such as those of locusts (*Anacridium, Locusta, Schistocerca* spp.) and gypsy-moth (*Lymantria dispar*) and the huge flocks of small grain-eating birds which congregate in cereal-growing regions. A specific instance of 'ruderal' dominance in birds is that described by Orians and Collier (1963) who observed that tricoloured blackbirds (*Agelaius tricolor*) through force of numbers are able to exclude a larger species, the redwinged blackbird (*Agelaius phoeniceus*).

'Competitive' dominance appears to be analogous in certain respects to the process of social dominance recognized by animal ecologists such as Allee (1938), Elton and Miller (1954), Park (1954), Miller (1967), Morse (1971), and Wilson (1971). Social dominance has been described by Morse (1974) as 'the priority of access to resources that results from successful attacks, fights, chases, or supplanting actions, present or past.' The appropriateness of this definition will not be lost on anyone who has kept a bird table or who has observed the feeding behaviour of species such as the starling (*Sturnus vulgaris*) or herring gull (*Larus argentatus*). As in the case of competitive dominance in plants, social dominance in animals is achieved at the cost of a high rate of reinvestment of captured resources and it is most characteristic of animals exploiting productive habitats.

Parallels between animals and 'stress-tolerant' dominants are rather more tenuous. Nevertheless, it is interesting to note that, in mature ecosystems, dominant effects of large mammalian herbivores upon smaller animals appear to be exerted not through direct social interactions but through resource depletion and habitat destruction arising from the presence of a high density of large animals. In a recent paper, Janzen (1976) has suggested that in countries such as Kenya and Uganda the scarcity of reptiles may be related to the dominant effect of large mammals. He postulates that 'through intensive grazing, browsing, and trampling, especially near water-courses during severe dry seasons, large herbivores should greatly reduce the cover available for reptiles, the small vertebrate prey of snakes, and the insects available to reptiles.' Janzen also points out that where there is a large biomass of herbivores there is likely to be a high density of carnivores which may exert an 'incidental depressant impact on the reptile biomass.'

Chapter 5

Succession

INTRODUCTION

From the work of numerous ecologists, most notably that of Clements (1916) and Watt (1947), it is now generally accepted that most types of vegetation are subject to temporal changes both in species composition and in the relative importance of constituent life-forms. These changes may be classified broadly into two kinds: successional and cyclical. During successional change there is a progressive alteration in the structure and species composition of the vegetation, whereas in cyclical change similar vegetation types recur in the same place at various intervals of time. In the case of succession, a further distinction may be drawn between the changes which occur during the colonization of a new and skeletal habitat initially lacking in soil and vegetation (primary succession) and those which characterize the much more common circumstance in which succession is a feature of the process of recolonization of a disturbed habitat (secondary succession).

More recent investigations and reviews (e.g. Watt, 1960; Whittaker, 1975; Odum, 1971; Drury and Nisbet, 1973; Horn, 1974; Fox, 1977) have progressed beyond the descriptive phase and have begun the task of identifying the mechanisms which cause various types of vegetation change. The objective in this chapter will be to augment these studies by considering the involvement of plant strategies in successional and cyclical vegetation changes. Although, initially, attention will be mainly confined to strategies of the established phase, the last section will be concerned with the role of regenerative strategies.

Some important clues to the role of strategies in successional change may be found in the relationships between the primary strategies and life-forms summarized in Figure 19. From this figure it is apparent that whereas the ruderal strategy comprises a fairly homogeneous group of ephemeral plants with many similarities in life-history and ecology, the competitors consist of a wide range of plant forms including perennial herbs, shrubs, and trees. However, the most remarkable conjunctions in plant form and ecology occur within the stress-tolerators where we find such apparently diverse organisms as lichens and certain forest trees.

In order to explain how this morphological variety among the ranks of the competitors and stress-tolerators is a key to certain aspects of the mechanism of vegetation succession it is necessary to compare the sequence of strategies and life-forms which appear in secondary successions in productive and unproductive environments.

SECONDARY SUCCESSION IN PRODUCTIVE ENVIRONMENTS

In localities where the soils and climate are conducive to high productivity there is a strong tendency for undisturbed vegetation to become dominated by trees, and clearance of woodland is followed, in the absence of further major disturbance, by a characteristic process of recolonization in which annuals, perennial herbs, shrubs, and trees are successively represented in the vegetation.

Succession in productive habitats is a complex phenomenon reflecting the relative dispersal efficiencies of plants (see pages 51 and 115) and progressive modification of soil and micro-climate by the changing vegetation. There can be little doubt also that competitive-dominance plays a major part in the process by which the annuals are displaced by perennial herbs. However, effects of dominance during the later stages of succession are more varied and complex and require careful analysis.

It is tempting to explain the sequence perennial herbs → shrubs → trees simply in terms of the longer period required for the development of the elevated leaf canopies of woody species. Reference has been made already to the comparatively slow growth-rates of shrubs and trees, an inevitable consequence of the expenditure of photosynthate upon supporting structures (trunks, branches, etc.) at the expense of leaf area. However, the possibility may be considered that the process of succession is further attenuated by the effects of perennial herbs upon seedlings and saplings of woody species. As explained on pages 131–2 the establishment of tree seedlings may be strongly inhibited by the presence of competitive herbs which, because of their capacity for lateral spread above and below ground, are better equipped, in the short-term, to dominate the vegetation. As an extension of this hypothesis, we may suppose that the tendency of shrubs to dominate the intermediate phase of vegetation succession in productive habitats is related to the branching form of the shoot which, in contrast to that of most trees, allows, at an early stage of growth, lateral spread of the leaf canopy and effective competition with herbs for light.

The woody species which appear early in the course of vegetation succession in productive habitats resemble in several respects the competitive herbs which they usurp. Their competitive characteristics include rapid (i.e. in comparison with other trees) rates of dry matter production (Plate 27), continuous stem extension and leaf production during the growing season, and rapid phenotypic adjustments in leaf area and shoot morphology in response to shade. These features are particularly conspicuous and have been documented (see page 26) among the deciduous trees and shrubs such as *Rhus glabra*, *Ailanthus altissima*,

Betula populifolia, Populus tremuloides, and *Liriodendron tulipifera* which occur at the early phase in reafforestation of disturbed woodland, abandoned pastures, and arable fields on the eastern seaboard of North America. Species which appear to play an equivalent role in Europe include *Sambucus nigra* and species of *Salix, Populus,* and *Betula*.

An additional characteristic of the competitive trees and shrubs is their intolerance of deep shade: as the tree canopy closes, seedling establishment in these species diminishes and many of the smaller saplings become severely etiolated and die. By this stage in vegetation succession it is usually evident that changes have also occurred in the composition of the herb layer. Although the competitive herbs are still represented, most of them are extremely reduced in vigour and tend to be replaced by evergreen, shade-tolerant herbs.*

As the larger of the competitive trees reach maturity most of the shrubs succumb to shade or reach the end of their life-span and become senescent. A new element then becomes prominent, consisting of trees which are, to varying extents, shade-tolerant. Seedlings of many of the shade-tolerant trees may have been present in relatively small numbers since an early stage of the succession, but because of their slow growth-rates they have been outstripped by the competitive trees and shrubs and have remained inconspicuous. The shade-tolerant trees differ from their predecessors in characteristics of both the established and regenerative phases of their life-cycles; these differences have been reviewed at some length in earlier chapters of this book. Many possess large seeds which, by providing an initial source of energy and structural materials, buffer the seedlings against the stresses experienced during establishment beneath a closed canopy of herbs, shrubs, or trees.

Although the shade-tolerant trees of productive habitats are slow-growing, they include certain species which have a relatively long life-span, attain a large size, and eventually, in the later stages of succession, become the dominant component of the forest. Over a large area of Central Europe, beech (*Fagus sylvatica*) plays this role, whilst in Australasia evergreens belonging to the genus *Nothofagus* are often the dominant trees in the final phase of succession in initially productive habitats.

A problem of great theoretical interest and practical significance concerns the mechanism whereby these climax trees attain their dominant status. Some authorities, such as Walter (1973), have attributed success to their competitive ability. However, characteristics of many of these trees, such as their infrequent fruiting (McClure, 1966; Burgess, 1972; Harper and White, 1974), suggest that they should be classified as stress-tolerant competitors (page 67) rather than competitors. Consideration of the circumstances prevailing in the later stages of succession (e.g. Hellmers, 1964; Whittaker, 1966; Odum, 1969) leads one to suspect that stress is likely to be a major selective force in climax forest. A consequence of the development of mature forest is the imposition of severe

*In temperate regions such as the British Isles there is usually also an incursion by vernal herbs such as *Ranunculus ficaria* and *Endymion non-scriptus*.

stress by the vegetation itself. Stress arises not only as a result of shading by a dense, continuous leaf canopy (Horn, 1971) and the declining ratio of photosynthetic production to respiratory burden (Lindeman, 1942; Odum, 1971), but also because a high proportion of the mineral nutrients present in the habitat are sequestered in the plant biomass (Jordan and Kline, 1972; Odum, 1971) and are conserved by very efficient and closed (*sensu* Odum, 1971) systems of mineral nutrient cycling.

SECONDARY SUCCESSION IN UNPRODUCTIVE ENVIRONMENTS

Where the productivity of the habitat is low, the role of ruderal plants and competitors in secondary succession is much contracted, and stress-tolerant herbs, shrubs, and trees become relatively important at an earlier stage. The growth form and identity of the climax species varies according to the nature and intensity of the stresses occurring in the habitat. Where the degree of stress is moderate, such as in semi-arid regions or on rather shallow nutrient-deficient soils, the climax vegetation is often composed of sclerophyllous shrubs and small, relatively slow-growing trees, typical examples of which are the North American species *Pinus aristata* (Currey, 1965; Ferguson, 1968) and *Quercus gambellii* (Cottam *et al*., 1959). It is not unusual for the canopy provided by such woody species to be sufficiently discontinuous to allow the presence of an understorey of stress-tolerant herbs (Siccama, Bormann, and Likens, 1970; Arno and Habeck, 1972; van Steenis,1972; Westman, 1975). Under more severe stress, such as that occurring in arctic and alpine habitats, ruderal and competitive species may be totally excluded and here both primary and secondary successions may simply involve colonization by lichens, certain bryophytes, small herbs, and dwarf shrubs (Muller, 1940; Shreve, 1942; Viereck, 1966).

A MODEL OF VEGETATION SUCCESSION

From the preceding description it is evident that a major factor determining the role of strategies in vegetation succession is the potential productivity of the habitat. This effect may be summarized in the form of a simple model (Figure 42).

The model is basically the same as that previously considered in Chapter 2 (Figure 13) and consists of an equilateral triangle in which variation in the relative importance of competition, stress, and disturbance as determinants of the vegetation is indicated by three sets of contours. At their respective corners of the triangle, competitors, stress-tolerators, and ruderals become the exclusive constituents of the vegetation. The areas within the triangle corresponding to the strategic ranges of selected life-forms and taxa are indicated in Figure 18.

The curves S_1, S_2, and S_3 in Figure 42a describe, respectively, the paths of succession in conditions of high, moderate, and low potential productivity, whilst the circles superimposed on the curves represent the relative size of the plant biomass at each stage of the succession. By reference to Figure 18 the

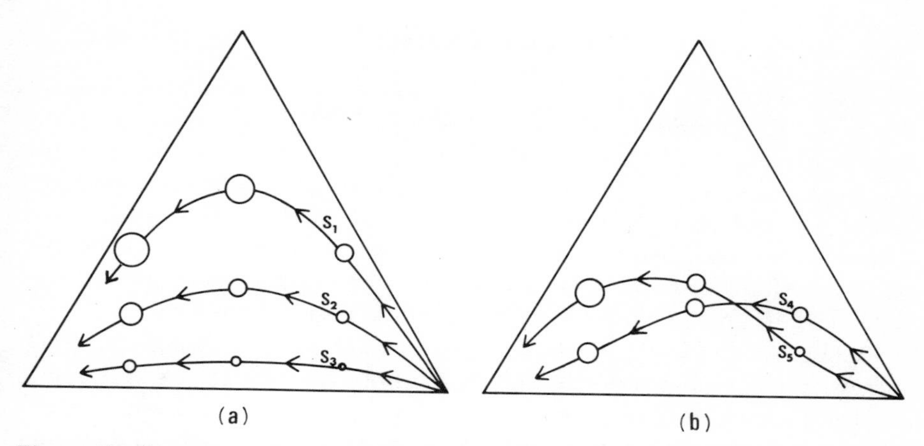

Figure 42 Diagrams representing the path of vegetation succession (a) under conditions of high (S_1), moderate (S_2), and low (S_3) potential productivity, and (b) under conditions of increasing (S_4) and decreasing (S_5) potential productivity. The size of the plant biomass at each stage of succession is indicated by the circles. For the distribution of strategies and life-forms within the triangle, see Figures 13 and 19 respectively. (Reproduced from *American Naturalist*, **111**, by permission of the University of Chicago Press. © 1977. The University of Chicago Press.)

curves may be used to predict the sequence of dominant life-forms in each succession.

In the most productive habitat, the course of succession (S_1) is characterized by a middle phase of intense competition in which first competitive herbs then competitive, shade-intolerant shrubs and trees dominate the vegetation. This is followed by a terminal phase in which stress-tolerance becomes progressively more important as shading and nutrient stress coincide with the development of a large plant biomass dominated by large, long-lived forest trees.

Where succession occurs in less productive habitats (S_2) the appearance of highly competitive species is prevented by the earlier onset of resource depletion and the stress-tolerant phase is associated with dominance by smaller slow-growing trees and shrubs in various vegetation types of lower biomass than those occurring in S_1.

In an unproductive habitat, the plant biomass remains low and the path of succession (S_3) moves directly from the ruderal to the stress-tolerant phase.

The curves of S_1–S_3 refer to succession under conditions of fixed potential productivity, a situation which is unlikely to occur in nature. In most habitats, the process is associated with a progressive gain or loss in potential productivity. Increase in productivity, for example, may arise from nitrogen fixation or input of agricultural fertilizer. Conversely, losses of mineral nutrients may result from cropping or leaching of the soil by rainfall. In Figure 42b curves have been drawn to describe hypothetical pathways of succession under conditions of increasing (S_4) and decreasing (S_5) potential productivity.

CLIMAX AND PROCLIMAX

In certain habitats the process of vegetation succession terminates in vegetation of a stable and predictable type—the so-called climax vegetation. In the most extreme type of climax vegetation, one species remains indefinitely as the dominant, and senescent individuals are replaced by members of the same species. A good example here is provided by the Central European forests dominated by beech (*Fagus sylvatica*). In many environments, however, there is no such well-defined end-point to succession and two or more species may co-exist in dynamic mixtures such as those described in Chapter 6.

Throughout the world, there are numerous examples of habitats in which succession has been arrested by severe forms of disturbance such as burning, grazing, mowing, or ploughing. In a given geographical region each form of disturbance when applied repeatedly in an orderly fashion results in a characteristic vegetation type usually described as a proclimax. Where the intensity of damage falls short of complete destruction of the established vegetation (e.g. grazing, mowing, burning) the result, in a potentially productive habitat, is to arrest succession at the stage where the the dominant species are competitors. Under the most severe forms of constant disturbance, e.g. arable fields (page 42), succession does not progress beyond the ruderal phase.

It has long been recognized (e.g. Tansley, 1939) that the effect of orderly disturbance is not merely to arrest succession but to change the character of the vegetation. This arises because certain forms of disturbance when constantly applied tend to bring into prominence species which play a minor role or may not even be represented in the 'normal' (i.e. uninterrupted) course of succession. Examples of plants which are encouraged by constant application of particular types of disturbance are provided by the many genera of turf grasses (e.g. *Agrostis, Festuca, Lolium, Poa*) in which the established plant is adapted to regular grazing or mowing. Further examples include the poisonous or distasteful herbs (e.g. *Senecio jacobaea, Ranunculus acris*) which are favoured by persistent overgrazing and the shrub *Calluna vulgaris* which flourishes in heathlands managed by rotational burning.

It would be misleading, however, to interpret the ecology of proclimax species simply in terms of the adaptation of the established phase of the life-cycle to particular forms of regular disturbance. As we shall see under the next heading, some of the most consistent and important differences between successional and proclimax plants relate to their regenerative strategies.

REGENERATIVE STRATEGIES AND VEGETATION DYNAMICS

Already in the concluding sections of Chapter 3 (pages 112–116), an attempt has been made to examine the ecological significance of particular regenerative strategies in various types of life-history. Here, therefore, only a brief reiteration of certain points is required in order to relate this information to the processes of vegetation change considered in this chapter.

Regenerative strategies in secondary successions

Where succession is initiated by localized disturbance of mature forest, the initial phase of recolonization usually involves those herbs, shrubs, and trees which produce an abundance of small, wind-dispersed seeds. By the time that succession has reached the stage where there is a dense vegetation cover (often as early as the second year following disturbance) conditions no longer favour the establishment of small, relatively fast-growing seedlings, and survival of the ephemeral element among the primary colonists depends critically upon its ability for renewed dispersal to freshly-disturbed sites.

A longer occupation is possible, however, among the perennial herbs, shrubs, and trees, in some of which regeneration by vegetative expansion allows consolidation of the more vigorous of the genotypes surviving from the original population of colonizing seedlings. Simultaneously or subsequently, these same plants release large quantities of seeds which facilitate establishment in new disturbed sites. The significance of this 'escape mechanism' in the primary colonists becomes apparent as succession proceeds to a stage where resource limitation exercises the dominant selective force with respect to strategies of both the regenerative and the established phase. At this late stage of succession, the most successful regenerative strategies appear to be vegetative expansion and that involving a bank of persistent seedlings, recruitment from which occurs mainly in response to mortalities among the established populations.

This description of the role of regenerative strategies in secondary succession is idealized and is based very largely upon patterns observed in the temperate forest environment. Quite clearly, more complex sequences will occur in circumstances where succession is interrupted more frequently; here one result may be to allow the persistence of species regenerating by wind-dispersed seeds. However, where disturbance is both frequent and extensive, e.g. in forests with a long history of damage by fire, grazing animals, or wind-falls, it is not unusual to find represented in the flora species in which regeneration involves a persistent seed bank (see pages 87–98).

Regenerative strategies in proclimax vegetation

The survival of proclimax vegetation depends upon continuous and, in some cases, severe disturbance and it is not surprising therefore to find that the two most common regenerative strategies (seasonal regeneration in gaps and regeneration from a persistent seed bank) in plants of arable land, pastures, meadows, and heathlands both confer the ability for *in situ* expansion or re-establishment of plant populations.

In circumstances where disturbance is a constant but seasonal effect of climate or is a seasonally-predictable biotic effect, the predominant regenerative strategy is usually that involving the production of a crop of seedlings or vegetative offspring at the time of year most propitious for re-establishment in the denuded areas. Where disturbance is less predictable there is a strong tendency for regeneration from a seed bank to become the most important strategy. The advantage of this form of regeneration is especially apparent where the vegetation is subject to intermittent disturbance. In such an environment the seed bank allows persistence of the population not only during periods of frequent disturbance but also over intervals in which the vegetation is relatively undisturbed and opportunities both for seedling eestablishment and even perhaps survival of the established plant are restricted.

Although the two most familiar strategies encountered in proclimax vegetation involve seed banks or seasonal regeneration, other types are not excluded. Where the intensity of disturbance is relatively low, regeneration by vegetative expansion is often exceedingly common among proclimax herbs and shrubs (page 81) and the large populations of composites such as *Taraxacum officinale*, *Hypochoeris radicata*, and *Leontodon hispidus*, in temperate grasslands, provide conspicuous evidence that the production of numerous wind-dispersed seeds has remained a viable alternative strategy in some forms of proclimax vegetation.

CONCLUSIONS

The general conclusion which may be drawn from this chapter is that processes of change in the structure and composition of vegetation can be interpreted as a function of the strategies of the component plant populations. Moreover, the relationships between established and regenerative strategies, life-forms, and habitat productivity provide a succinct basis upon which to explain the various sequences of plants observed in secondary successions. It would appear also that knowledge of the regenerative strategies of plants is particularly relevant to an understanding of the difference between successional and proclimax vegetation.

It is now pertinent to consider whether by the recognition of strategies it is possible to analyse other features of vegetation change which are of interest to plant ecologists. These include differences between early and late successional stages in rate of floristic change, in vegetation stability and in species density.

Rates of floristic change during secondary successions

There appears to be general agreement between past and present authorities (e.g. Clements, 1916; Tansley, 1939; Margalef, 1968; Whittaker, 1975) that the progress of secondary succession is usually marked by a progressive decline in the rate of floristic change, and from certain studies such as that of Hanes (1971) it may be inferred that relatively slow rates of vegetation change characterize secondary succession in unproductive habitats.

Especially in highly productive environments, a rapid turnover of populations and species occurs during the phase of initial colonization and vegetation development. This is followed by slower rates of species replacement once the ruderals have given way to competitors, stress-tolerant competitors, or stress-tolerators, the majority of which are comparatively long-lived. Other mechanisms which tend to limit the rate of vegetation change late in succession are the increasing scarcity of conditions suitable for seedling establishment and the low seed production and slow dispersal rates of many late succession species.

Changes in vegetation stability and species density during secondary successions

It has been pointed out by Margalef (1963a, b, 1968) and Horn (1974) that some ambiguities may arise when the word 'stability' is used to characterize successional stages. In order to avoid confusion it is necessary to distinguish between *inertia*, i.e. resistance of the undisturbed vegetation to change (increasing during succession) and *resilience*, i.e. ability to recover rapidly from disturbance (decreasing during succession). The resilience of late successional vegetation is diminished not only by the slow rates of recolonization and growth of its component plants but also by the fact that destruction of all or part of the biomass of a mature ecosystem may lead to irreversible changes in mineral nutrient status and soil structure such as those frequently observed after the felling of tropical rain-forest.

During secondary succession there are various developments, each of which is capable of influencing species density (i.e. number of species per unit area). These include increasing modification of the soil and micro-climate by the vegetation, increasing interaction between plants, declining frequency of seedling establishment, and increasing stratification of the vegetation. Certain of these changes tend to increase species density whilst others have the opposite effect. It is clear also that patterns of change in the species density of one layer of the vegetation may be out of phase with changes in another. Both Margalef (1963b) and Odum (1971) have recognized that, during the intermediate stages of forest succession, species density in the shrub and tree canopy may continue to increase at a time when effects of dominance by the woody plants are causing a progressive reduction in species density in the herb layer.

A number of ecologists, including Bertalanaffy (1950), MacArthur (1955), Odum and Pinkerton (1955), Bray (1958), and Margalef (1963b, 1968), have suggested that the progress of succession is marked by increasing inertia, reduced dominance, and increasing species density. Whilst there seems to be no reason to doubt the general validity of this hypothesis it is necessary to bear in mind that in modern landscapes two new factors are operating which drastically limit rates of succession and diversification.

The first of these limiting factors arises from the trend in most parts of the world towards dissected and depauperate floras. This change arises not only from increasing urbanization but also from intensive methods of agriculture and forestry. One effect of such changes may be to restrict the distribution of late successional plants to 'islands', emigration from which is limited by their characteristically low reproductive output and dispersal efficiencies. In Northern England this phenomenon may be observed in extreme form in derelict industrial landscapes such as those in the southern parts of Lancashire and Yorkshire. Here refuges containing late-successional trees, shrubs, herbs, and cryptogams are extremely localized and in the intervening areas of wasteland succession may remain suspended indefinitely at the visually monotonous stage of competitive-dominance by a relatively small number of mobile early-successional herbs and shrubs. In this type of situation it seems inevitable that landscape reclamation and nature conservation must involve procedures whereby succession is accelerated and diversity created through deliberate introductions of under-dispersed plants (Grime, 1972; Bradshaw, 1977).

A second development which may affect the course of vegetation succession in intensively-exploited landscapes is that of eutrophication. As we have seen earlier in this chapter, there is reason to suspect that many of the plants which occur in mature ecosystems of high floristic diversity are adapted to conditions of resource limitation. Where the effect of agriculture is to release mineral nutrients such as phosphorus and nitrogen into natural habitats the effect may be to prevent resource depletion and to arrest succession at the stage of competitive dominance. This may not be an exclusively modern phenomenon. Where vegetation on alluvial terraces is subject to mineral nutrient inputs from flood water it seems likely that conditions may arise which favour continuous occupation of the habitat by relatively fast-growing trees, shrubs, and herbs.

Chapter 6

Co-existence

INTRODUCTION

As progress is made in understanding the primary mechanism of vegetation, the possibility arises that some generalizations can be attempted with regard to the processes which control the number of plant species per unit area of vegetation. Species density is, of course, affected by many factors, and these vary in importance according to vegetation type and sample size. Where the area of the sample is large, a high species density may arise from the fact that the sample includes a mosaic of quite different environment and vegetation types corresponding to topographical variation, gradients in soil and climate or patterns of vegetation management. However, as the size of the sample is reduced, major discontinuities in environment become less important and, in very small samples, the determinants of species density are those which allow plants of different biology to establish and survive in close proximity. It is these more subtle mechanisms which provide the subject of this chapter.

CO-EXISTENCE IN HERBACEOUS VEGETATION

In the preceding analysis of dominance (Chapter 4) reference was made to the tendency of larger herbaceous plants to suppress the growth and regeneration of smaller neighbours and it was concluded that, especially in fertile environments, a trend towards monoculture is frequently observed providing that the biomass remains relatively undisturbed. We may suspect therefore that one of the preconditions for the co-existence of species in herbaceous vegetation is that factors are present which limit the expression of dominance. This limitation may operate through stress or disturbance or by a combination of the two, and its effect is usually to debilitate the potential dominants and to allow plants of smaller stature to regenerate and to co-exist with them.

The influence which stress and disturbance may exert (together and in isolation) upon the degree of dominance is illustrated in Figure 43, which describes

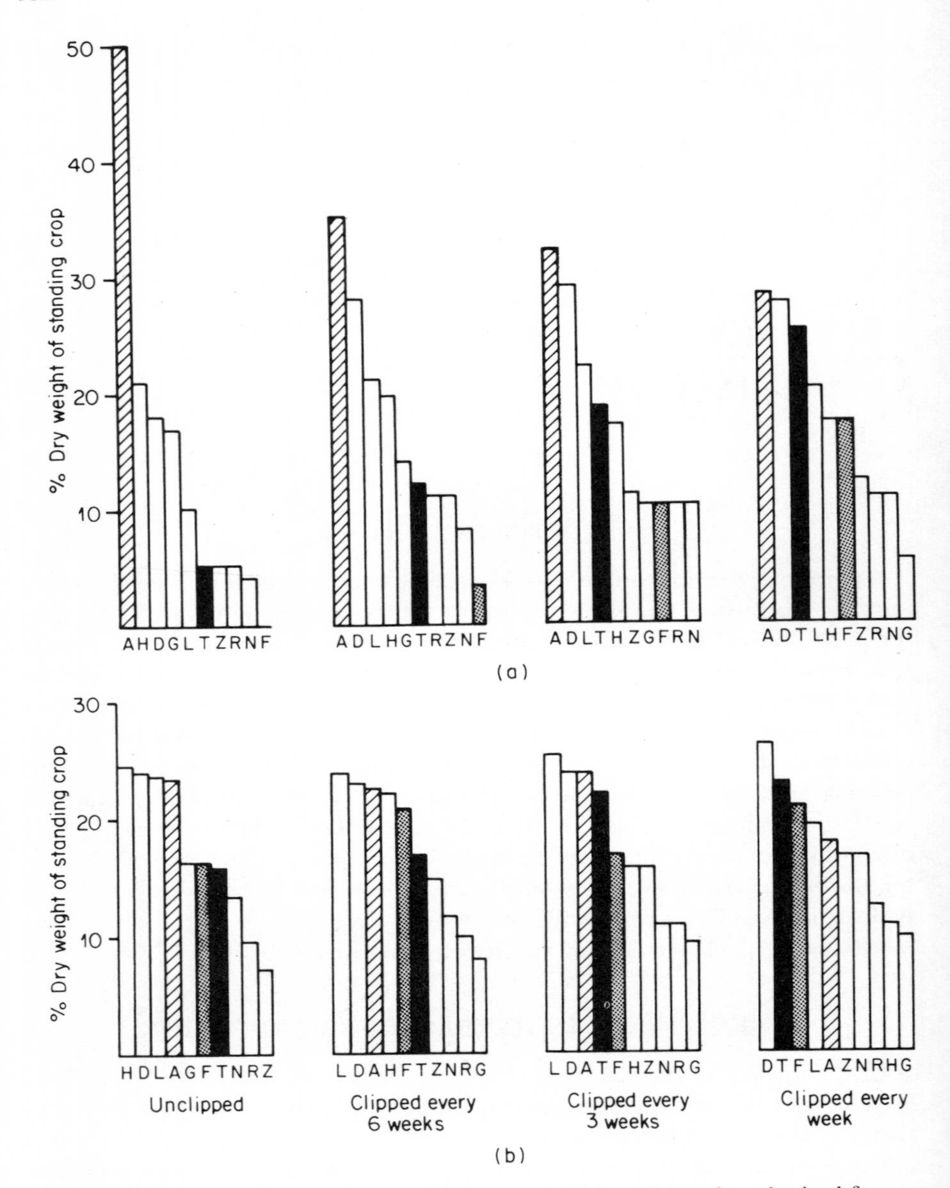

Figure 43 Histograms comparing the species composition of a turf synthesized from a standardized seed mixture and subjected to (a) high fertility (170 mg/l N) and (b) low fertility (5 mg/l N) and four intensities of defoliation. The experiment was conducted in 36 l containers over a period of 7 months and the results are expressed as percentages of standing crop transformed to angles. Key to species: A, *Arrhenatherum elatius;* H, *Holcus lanatus;* D, *Dactylis glomerata*; G, *Agropyron repens*; L, *Lolium perenne*; T, *Agrostis tenuis*; Z., *Bromus erectus*; R, *Festuca rubra*; N, *Anthoxanthum odoratum*; F, *Festuca ovina* (Mahmoud, 1973).

the result of an experiment in which stress (represented by nitrogen deficiency) and disturbance (represented by mowing) were applied to a turf derived from a standardized seed mixture of common grasses. The data confirm that the effect of both nitrogen stress and frequent mowing was to suppress *Arrhenatherum elatius*, the dominant under fertile, undisturbed conditions. It is interesting to note, moreover, that the most equitable distribution of shoot dry matter between the component species occurred in the treatment involving both nitrogen stress *and* a high frequency of defoliation.

From both field observations and experiments there is a wealth of additional evidence supporting the hypothesis that co-existence between a large number of herbaceous species (i.e. high species density) is associated with the debilitation of potential dominants by effects of environment or management. The addition of mineral fertilizers to nutrient-deficient vegetation has been found to cause an expansion in species of large stature, with a coincident reduction in species density (Thurston, 1969; Willis, 1963; Jeffrey, 1971). Conversely, treatment of productive vegetation with the growth retardant maleic hydrazide tends to suppress potential dominants and to increase species density (Yemm and Willis, 1962); a similar effect has been achieved by grazing and herbage removal (Singh, 1968; Singh and Misra, 1969) and by treatment of annual communities with short term doses of gamma radiation (Daniel and Platt, 1968).

Measurements of species density in a wide range of natural and semi-natural vegetation types in Northern England (Grime 1973c) has provided circumstantial evidence of the importance of factors such as grazing, mowing, burning, and trampling in the maintenance of high species densities in certain habitats. From Figure 44 it is clear that types of perennial herbaceous vegetation subject to disturbance (e.g. pastures and meadows) in general display species densities higher than those of unmanaged habitats.

The impact of management is most apparent when species densities in derelict sheep pastures of the Derbyshire dales are compared with those from neighbouring dales in which grazing has been maintained to the present day. From comparisons such as that illustrated in Figure 45 there is evidence that the beneficial effect of continued grazing upon species density extends over the full spectrum of grasslands associated with the soil pH range 4.0–8.0. A similar effect of disturbance upon dominance and species density has been observed in North America in areas of tall-grass prairie exploited by prairie dogs (Koford, 1958) and badgers (Platt, 1975).

From several sources, therefore, we have evidence substantiating the hypothesis that co-existence is encouraged by effects of stress and disturbance upon vegetation. However, in order to obtain a more complete assessment of the role of these factors upon species density it is necessary to consider the rather different circumstances which arise when the intensities of stress and/or disturbance become severe. Further scrutiny of the measurements of species density in Figure 45 reveals that even in the mild temperate climate of Northern England, habitats may be identified (acidic pastures, coal mine heaps, conifer

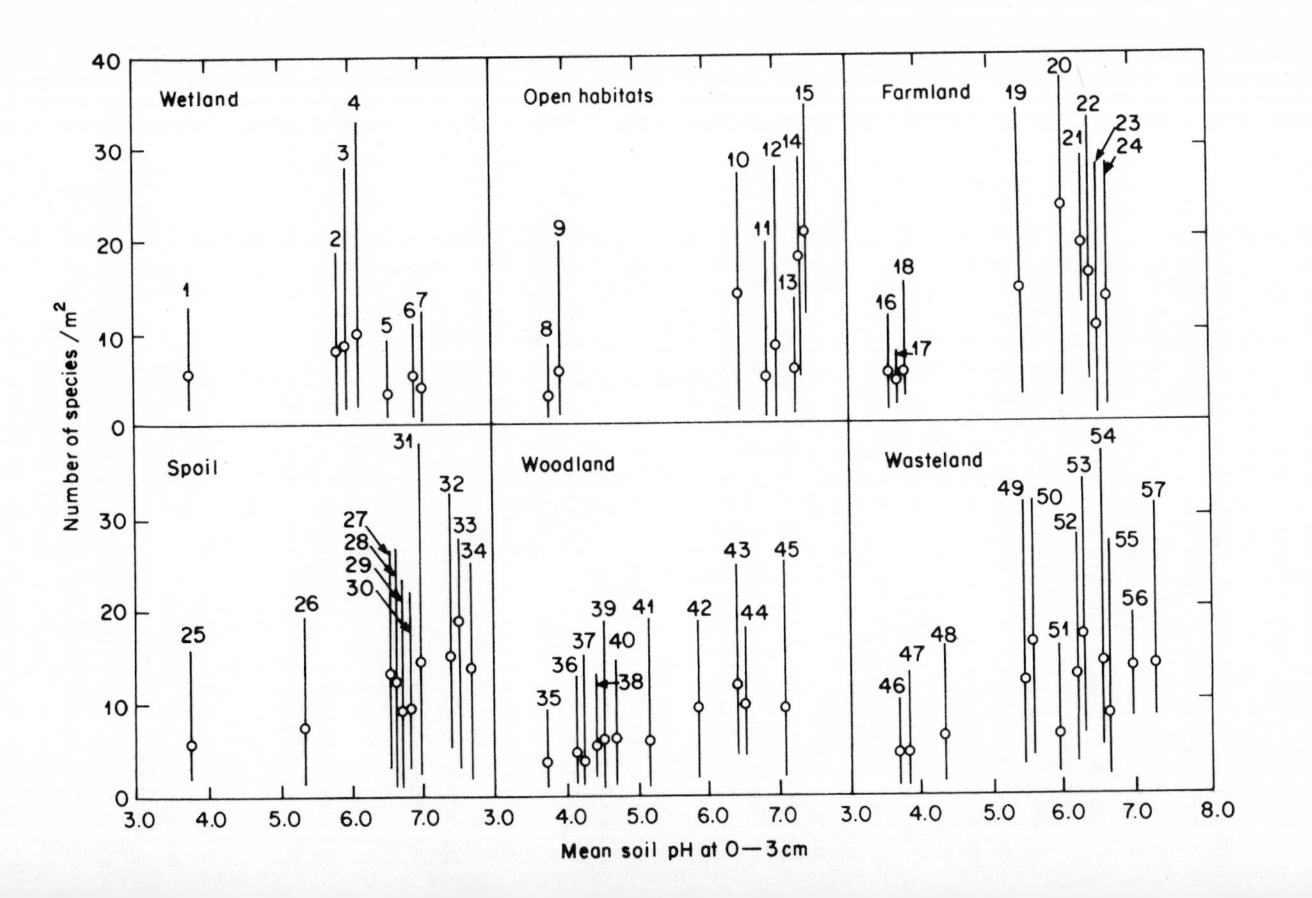
Wetland
Open habitats
Farmland
Spoil
Woodland
Wasteland
Number of species / m²
Mean soil pH at 0—3 cm

Figure 44 Mean and range in species density in the one square metre samples of vegetation examined in the major habitats present in an area of 900 square miles around Sheffield, Yorkshire. Bare ground was not sampled; hence the minimum value possible is one.

Key to habitats: 1, unshaded mire on noncalcareous strata, soil pH <4.1; 2, shaded mire, 3; unshaded mire on calcareous strata; 4, unshaded mire on noncalcareous strata, soil pH >4.0; 5, lakes, ponds, canals, ditches, depth of water <11 cm; 6, lakes, ponds, canals, ditches, depth of water >10 cm; 7, rivers and streams; 8, noncalcareous cliffs; 9, rock outcrops on noncalcareous strata; 10, consolidated limestone scree; 11, walls; 12, limestone cliffs; 13, unconsolidated limestone scree; 14, rock outcrops on Magnesian Limestone; 15, rock outcrops on Carboniferous Limestone; 16, unenclosed pasture on Millstone Grit; 17, unenclosed pasture on limestone, soil pH <4.1; 18, unenclosed pasture on Coal Measures; 19, enclosed pasture; 20, unenclosed pasture on limestone, soil pH >4.0; 21, permanent meadows; 22, fallow arable; 23, arable, crop broad-leaved; 24, cereal arable; 25, coal mine heap, soil pH <4.1; 26, coal mine heap, soil pH >4.0; 27, lead mine waste, discontinuous vegetation cover; 28, cinders; 29, manure and sewage residues; 30, lead mine waste, continuous vegetation cover; 31, heaps of mineral soil (building sites, etc.); 32, spoil heaps in limestone quarries (Carboniferous Limestone); 33, spoil heaps in limestone quarries (Magnesian Limestone); 34, brick and mortar rubble; 35, plantations of broad-leaved trees, soil pH <4.1; 36, woodland on Bunter Sandstone; 37, coniferous plantations; 38, scrub on noncalcareous strata; 39, woodland on Millstone Grit; 40, woodland on Coal Measures; 41, plantation of broad-leaved trees, soil pH <4.1; 42, woodland on Magnesian Limestone; 43, scrub on limestone; 44, hedgerows; 45, woodland on Carboniferous Limestone; 46, derelict grassland and heath on Bunter Sandstone, soil pH <4.1; 47, derelict grassland and heath on Millstone Grit; 48, derelict grassland and heath on Coal Measures; 49, derelict grassland and heath on Bunter Sandstone, soil pH >4.0; 50, derelict grassland and heath on Carboniferous Limestone; 51, shaded paths; 52, river banks, etc.; 53, woodland on Magnesian Limestone; 54, unshaded paths with incomplete vegetation cover; 55, unshaded paths with complete vegetation cover; 56, road verges, mown frequently; 57, road verges, mown annually or unmown. (Reproduced with permission from *J. Envir. Man.*, 1, 151–167. Copyright 1973 by Academic Press Inc. (London) Ltd.)

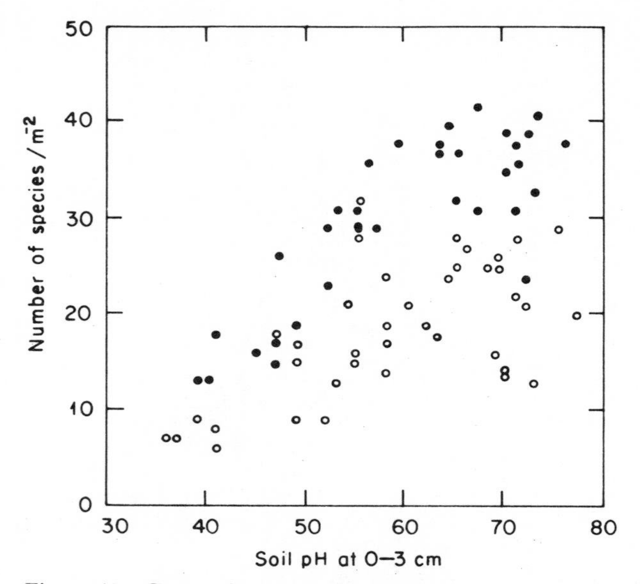

Figure 45 Comparison of species densities in unenclosed grassland in two Derbyshire dales differing in management (• Cressbrookdale; grazing by sheep and cattle. ○ Lathkilldale, ungrazed, burned sporadically). The samples refer to one square metre quadrats located at random. (Reproduced with permission from *J. Envir. Man.*, 1, 151–167. Copyright 1973 by Academic Press Inc. (London) Ltd.)

plantations, shaded paths) in which the intensities of stress and/or disturbance experienced by herbaceous vegetation are sufficient not only to eliminate potential dominants but to produce local environments which are inhospitable to all except a few specialized plants.

It is clear, therefore, that if we are to describe the mechanism controlling species density in herbaceous vegetation it will be necessary to recognize that the effects of stress and disturbance upon species density change according to their intensity. This changing relationship is the central feature of the simple model examined under the next heading.

A model describing the control of species density in herbaceous vegetation

In the 'hump-backed' model in Figure 46 it is suggested that the result of moderate intensities of either stress or disturbance (or both) is to increase species density by reducing the vigour of potential dominants, thus allowing subsidiary species to co-exist with them. At the most extreme inten-

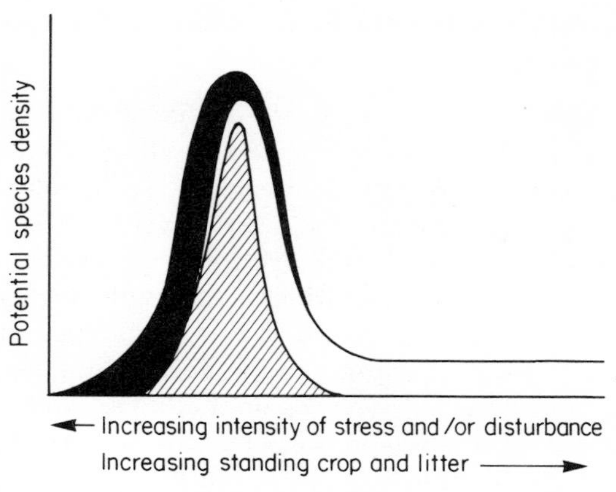

Figure 46 Model describing the impact of a gradient of increasing stress and/or disturbance upon the potential species density in herbaceous vegetation. □ potential dominants; ■ species or ecotypes, highly adapted to the prevailing form(s) of stress on disturbance; ▨ species which are neither potential dominants, nor highly adapted to stress or disturbance (Grime, 1973). (Reproduced by permission of Macmillan (Journals) Ltd.)

sities of stress and/or disturbance, however, the model suggests that species densities decline as conditions are created to which only a very small number of species are sufficiently adapted to survive.

This model is consistent with the observation of Odum (1963) that 'the greatest diversity occurs in the moderate or middle range of a physical gradient' and data conforming to the model have been obtained from transects along naturally-occurring gradients in stress and/or disturbance (Grime 1973a).

One of the most interesting and unusual examples of data conforming to this model is provided by the work of Woodwell and Rebuck (1967) who measured the changes in species density in the ground flora along a gradient of increasing exposure to gamma radiation. They conclude that 'the effects of irradiation do not appear to be simply a reduction in diversity but, at relatively low exposures where disturbance has been less, an increase in diversity as well.'

It would appear therefore that the model provides a convenient summary of some of the phenomena which influence species density, and this suggests that it may be used to predict or to explain the response of particular types of herbaceous vegetation to changes in management (see page 187). However, before exploring this possibility it is necessary to consider the quantitative aspect of the mechanism described by the model and to review some additional factors which influence species density.

Quantitative definition of the mechanism controlling species density in herbaceous vegetation

It has been argued (Al-Mufti *et al.*, 1977) that certain of the phenomena which influence species density in herbaceous vegetation (dominance, debilitation of dominants, incursion of subsidiary species, reduction of species density by extremes of stress and/or disturbance) are characteristic of particular ranges in the total dry weight of shoot biomass and litter. This hypothesis has been tested by examining the relationship which was obtained when species density was plotted against the annual maximum in standing crop + litter (both measured at the time of the maximum in the former) in a range of contrasted vegetation types, including tall herb communities, woodland floor communities, and grasslands, of common occurrence in Northern England. In order that the local reservoir of species (page 183) in the vicinity of the sites should not act as a major limiting factor upon species density, each was selected from an area of landscape containing a diversity of long-established vegetation types.

The resulting graph (Figure 47) conforms to the model and contains a corridor of high species density approximately in the range 350–750 g m^{-2}. From these results it would appear that above 750 g m^{-2} diversification tends to become restricted by dominance, whilst below 350 g m^{-2} species density is limited by the small number of species capable of surviving the severity of environmental stress and/or disturbance experienced in such extreme habitats.

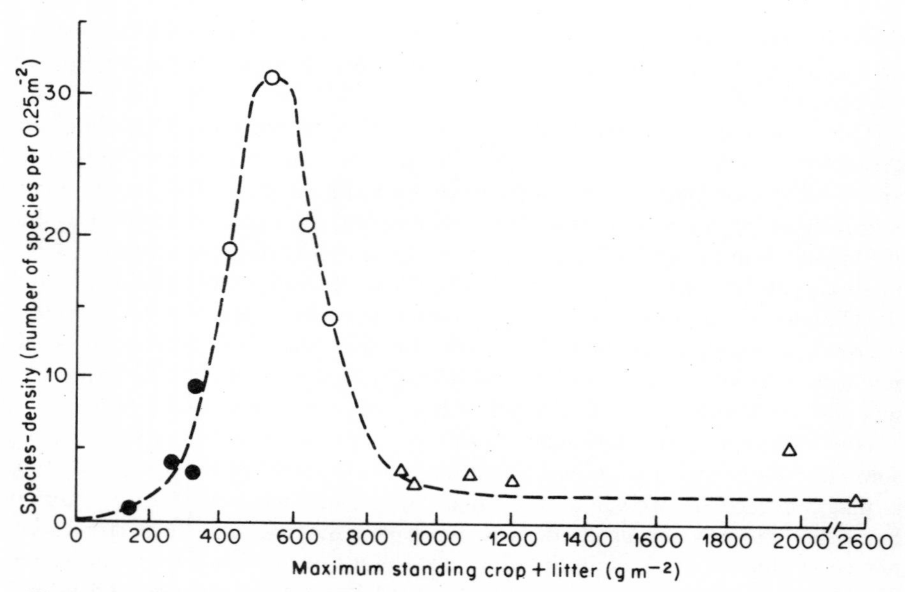

Figure 47 The relationship between maximum standing crop plus litter and species-density of herbs at fourteen sites in Northern England. ○ grasslands; ● woodlands; △ tall herbs (Al-Mufti *et al.*, 1977).

Within the corridor, however, we may suppose that vegetation composition is determined by an equilibrium between the conflicing forces of natural selection associated with moderate intensities of competition, stress, and/or disturbance, with the result that habitats are created in which co-existence is possible between a variety of species and genotypes.

Before considering in more detail the mechanisms whereby species of different habitat requirements may co-exist within 'corridor environments' it is necessary to make certain qualifying statements concerning the relationship between species density and the annual maximum in standing crop + litter. The precise form of the relationship in Figure 47 is clearly a function of sample size and will be affected also by factors peculiar to the sites and to the period of time at which this particular investigation was carried out. Moreover, for reasons which will be considered later (page 183), the maximum species densities attainable in corridor environments vary with latitude, regional location, and soil type.

Perhaps the most important qualification to be made with regard to the relationship in Figure 47 is that it is based upon measurements conducted on vegetation which has existed in its present state for some considerable time. In these circumstances we might expect, therefore, that the species densities which have been attained correspond to an equilibrium between the habitat condition prevailing at each site and the reservoir of species in its vicinity. In herbaceous vegetation which is of recent origin or is subject to marked changes in management we would not expect such a stable equilibrium. In certain habitats (e.g. road verges, semi-derelict hill pastures, the margins of hay-meadows) the maximum in standing crop + litter often varies to such an extent that conditions favourable to high species density in one year may be followed in the next by the development of a large and relatively undisturbed biomass. This effect is especially pronounced in the herb layers of woodlands where the density and persistence of deciduous tree litter at specific sites may vary considerably from year to year. Where vegetation is subject to such vicissitudes it is not to be expected that the maximum in biomass and litter measured in any one year will be predictably related to species density. An illustration of the extent to which, in such circumstances, definition of the pattern may be lost is provided in Figure 48 which shows the relationship between species density and estimations of the seasonal maxima in biomass + litter in samples drawn from a variety of habitats subject to erratic management.

A further qualification which must be made concerning the quantitative relationships in Figure 48 is that they apply to temperate conditions. From the investigations of Singh (1968) and Singh and Misra (1969) it is apparent that in tropical environments the corridor of high potential species density in herbaceous vegetation may extend to higher values of biomass + litter. This may be related to the fact that, in tropical grasslands, plant growth, flowering, senescence, and litter decomposition occur extremely rapidly and it is therefore possible for species with complementary phenologies to co-exist with minimal effects of dominance by the more productive components.

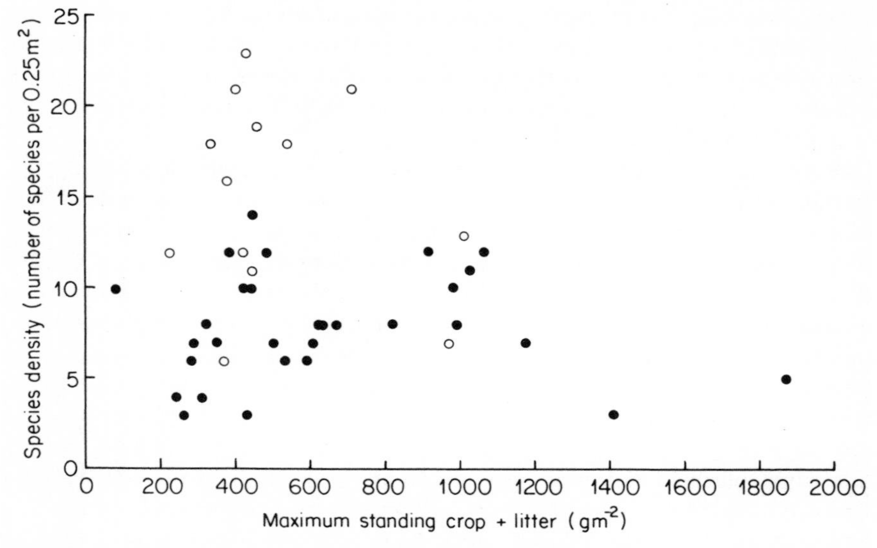

Figure 48 The relationship between estimates of the maximum standing crop plus litter and species-density of herbs in environments experiencing fluctuating patterns of vegetation management in Northern England. • road verges, subject to occasional mowing; ○ semi-derelict limestone pastures (Grime, Sydes, and Rodman, unpublished).

Co-existence related to spatial variation in environment

Horizontal variation

Even in apparently uniform terrain, most vegetation samples include a complex mosaic of micro-habitats. These arise from such factors as edaphic variation, interactions between micro-topography and climate, selective predation, local disturbance of the soil, and redistribution of nutrients by animals. To these must be added spatial differences in environment arising from the activity of the plants themselves. These include variation in nutrient availability (Snaydon, 1962), water supply, degree of shading, accumulation of litter and organic toxins (Abdul-Wahab and Rice, 1967), and modification of the soil micro-flora. It is the task of the plant ecologist to identify the circumstances in which this small-scale spatial variation is sufficient to initiate or to maintain conditions favourable to the co-existence of different species.

Many of the events which play a critical role in maintaining high species densities in herbaceous vegetation occur in the regenerative phase and are related to complementary mechanisms of establishment in vegetation gaps. The importance of gaps for the survival of seedlings and some vegetative offspring has been emphasized in Chapter 3 and from various sources there is conclusive evidence (e.g. Chippendale and Milton, 1934; Champness and Morris, 1948; Sarukhan, 1974; Thompson, 1977) that co-existence frequently occurs between species with very different regenerative strategies. As suggested by Grubb

(1977), it is not unreasonable to suspect that differences between neighbouring gaps in features such as size, soil texture, mineral nutrient status, desiccation, and degree of shading will be sufficient to encourage establishment by different species.

Although co-existence is frequently initiated by events occurring during the regenerative phase there are many instances where the pattern is reinforced by effects of spatial heterogeneity upon the vigour and survival of established plants. In the British Isles, some of the most conspicuous examples of this phenomenon occur in the herb layers of deciduous woodlands (Plate 16) where differences in topography and depth of tree litter often give rise to micro-environments exploited by species of contrasted size and growth form (Scurfield, 1953; Al-Mufti, 1978; Sydes, 1978). In North America also, circumstances have been described where co-existence between a variety of herbaceous plants may be related to the development of micro-habitats on the woodland floor. In one investigation described by Bratton (1976), for example, the range of micro-habitats and associated species included fallen logs (*Laportea canadensis*, *Sedum ternatum*, *Stellaria pubera*), the bases of large trees (*Osmorhiza longistylus*), the bases of small trees (*Dicentra cucullaria*, *D. canadensis*), and pockets of tree litter (*Phacelia fimbriata*).

In Britain, another vegetation type in which co-existence is related to spatial heterogeneity in the environment arises where sheep pastures are situated on shallow soils over fissured limestone (Plate 28, Figure 49a). In this habitat it is not uncommon to find four micro-habitats within areas of less than one square metre, each with a characteristic assemblage of herbaceous species. The first micro-habitat is that restricted to areas of extremely shallow soil and colonized by winter annuals such as *Arenaria serpyllifolia* and *Saxifraga tridactylites*. The second consists of patches where the soil is slightly deeper and the effects of summer desiccation are less severe, and here tussock grasses such as *Festuca ovina* and *Koeleria cristata* are located. Narrow deep fissures in the limestone are usually exploited by species such as *Poterium sanguisorba* and *Lotus corniculatus* which develop long, persistent tap roots. The fourth micro-habitat corresponds to the larger pockets of soil in which more productive species such as *Festuca rubra* and *Dactylis glomerata* are able to survive. Reference to Chapter 2, where the strategies exhibited by the four groups of plants just described are considered in more detail, reveals that in this example we are not concerned merely with an assemblage of basically similar grassland plants. On the contrary, we find that within the eight plants which have been cited as examples there are profound differences in morphology, phenology, and life-history and there is a strategic range including stress-tolerant ruderals (the small winter annuals), stress-tolerant competitors (*F. rubra* and *D. glomerata*) and two different types of 'C–S–R' strategists (tussock grasses and tap-rooted plants).

It is unusual for high species densities to be maintained in herbaceous vegetation simply as a result of environmental heterogeneity. As we shall see under the next heading, spatial and temporal differences in environment frequently coincide. It is also of the utmost importance, particularly in designing management

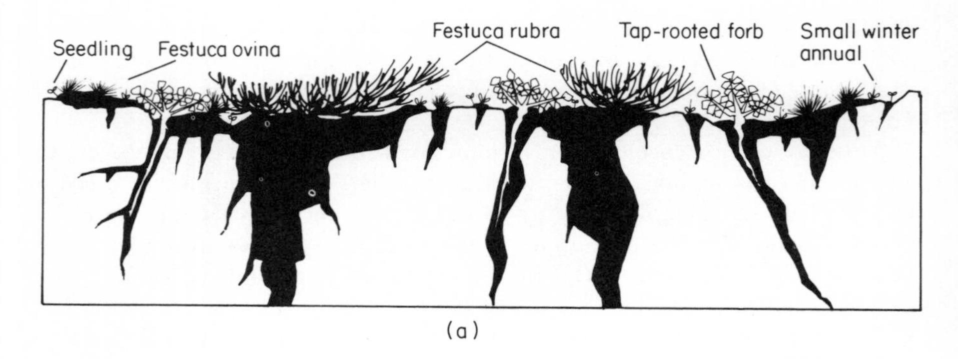

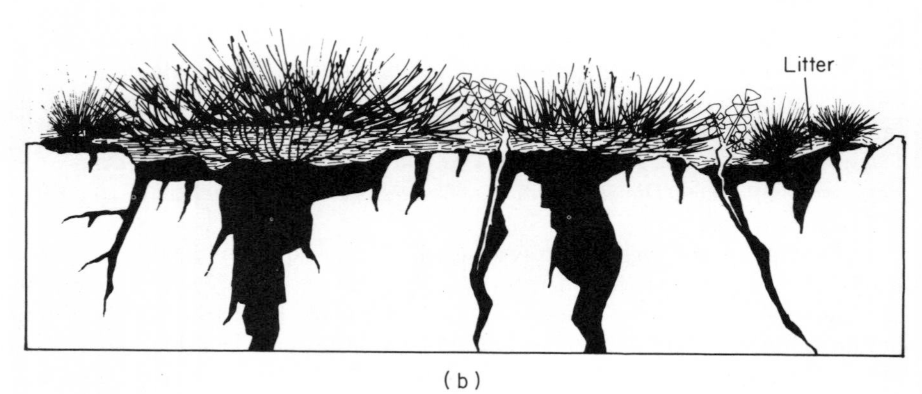

Figure 49 Diagram illustrating the vegetation structure associated with limestone outcrops of varying soil depth in North Derbyshire, England: (a) an outcrop subject to low intensity of grazing by sheep or cattle, (b) the same outcrop after several years of dereliction.

regimes for the purposes of nature conservation and amenity, to recognize that spatial variation in environment does not by itself provide a guarantee that high species densities will be maintained. This last point may be illustrated by reference to Figure 49b which depicts the kind of vegetation structure which has been observed to develop when limestone pastures over fissured limestone are allowed to become derelict. In the absence of grazing there is usually a marked increase in the vigour and lateral projection of the leaf canopy of the stress-tolerant competitors which are rooted in the larger pockets of soil. There is also a marked increase in accumulation of litter and, as the critical value of 750 g m^{-2} in maximum standing crop + litter (page 164) is approached, many of the areas of shallow soil exploited by the smaller plants become untenable. Reductions in species density are also detectable where the change in biomass is less dramatic than that illustrated in Figure 49b; in these circumstances we may suspect that the main impact of dereliction on species density arises from the deposition of leaf litter over the areas of bare soil which, prior to the cessation of grazing, provided opportunities for seedling establishment.

Vertical stratification

In productive, relatively undisturbed herbaceous vegetation, intense competition and the trend towards monoculture are associated with an extremely simple canopy structure in which the foliage of the dominant plants is concentrated in a dense, elevated layer (Plate 2) beneath which the frequency and vigour of smaller plants and seedlings are severely restricted by shading and deposition of litter (pages 125–129). However, where the productivity is lower and the canopy of the dominant plants is less complete (Figures 50 and 51, Plate 4) it is not unusual for the shaded stratum to include additional layers of leaves mainly composed of shade-tolerant herbs and bryophytes.

In moderately productive vegetation such as that illustrated in Plates 29 and 30, the layers in the leaf canopy are reconstructed annually by extension of new herbaceous shoots from the ground surface and by expansion of relatively fast-growing bryophytes on to the most recent increment of plant litter. A more stable form of vertical stratification may be observed in some of the less productive types of herbaceous vegetation, an example of which is dissected in Plate 31. In such damp, species-rich limestone pastures it is usually apparent that, even within short turf, there is considerable layering in the predominantly-evergreen canopy. In vegetation of this type there are often marked differences in the vertical distribution of the bryophytes, some of which (e.g. *Dicranum scoparium*) are in contact with the soil, whilst others (e.g. *Thuidium tamariscinum*) are attached to the standing litter of grasses and subshrubs or merely suspended in the herbaceous canopy (e.g. *Pseudoscleropodium purum*).

Co-existence related to temporal variation in environment

Seasonal variation

Close scrutiny of small samples of vegetation often reveals situations in which species of contrasted ecology are growing in intimate association and in which it is difficult to explain their co-existence in terms of spatial heterogeneity in environment. In many of these cases, it is apparent that the species concerned have different seasonal patterns of shoot expansion and flowering and are adapted to different parts of the annual climatic cycle.

A familiar example of the exploitation of temporal variation occurs in intensively-grazed lowland pastures in the British Isles where the sown grasses *Lolium perenne* and *L. multiflorum* exhibit peaks in biomass during the cool moist conditions of spring and autumn whilst White Clover (*Trifolium repens*), together with many of the naturally-invading indigenous species such as *Holcus lanatus* and *Agropyron repens*, have complementary phenological patterns. Similar differences in phenology have been recorded in tall-herb communities where the simplest types of co-existence (Figure 50) are those in which the huge midsummer standing crop of dominants such as *Urtica dioica*, *Chamaenerion angustifolium*, or *Petasites hybridus* is associated with minor ‘off-peak’ contributions from bryophytes or vernal herbs.

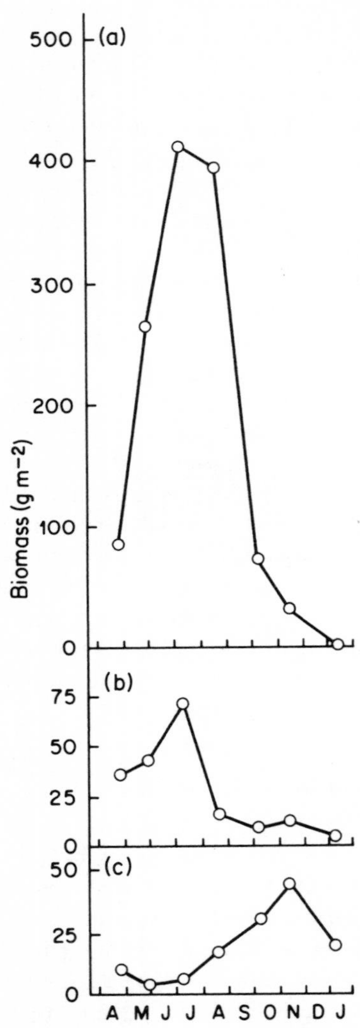

Figure 50 Seasonal changes in the shoot biomass of the main vegetation components in a stand of *Petasites hybridus* (Butterbur) situated on a lightly-shaded alluvial terrace in Northern England: (a) *Petasites hybridus*, (b) *Poa trivialis*, (c) Bryophytes (Al-Mufti *et al.*, 1977). (Reproduced by permission of Blackwell Scientific Publications Ltd.)

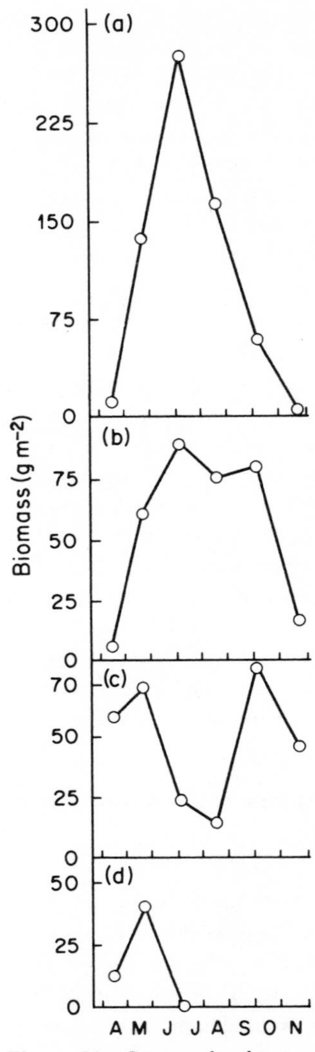

Figure 51 Seasonal changes in the shoot biomass of the main vegetation components in a stand of *Filipendula ulmaria* (Meadowsweet) situated on a damp, calcareous north-facing terrace in Northern England: (a) *Filipendula ulmaria*, (b) *Mercurialis perennis*, (c) Bryophytes, (d) *Anemone nemorosa* (Al-Mufti *et al.*, 1977). (Reproduced by permission of Blackwell Scientific Publications Ltd.)

A more complex example is illustrated in Figure 51 which describes some of the phenological changes in a stand of Meadow Sweet (*Filipendula ulmaria*). Here the well-defined summer peak in shoot biomass of the dominant herb (*F. ulmaria*) is combined with the broader distribution of *Mercurialis perennis*, a stress-tolerant competitor (see page 109) which is capable of surviving in the shaded stratum beneath *F. ulmaria*. The remaining contributors are the vernal herb *Anemone nemorosa* and the pleurocarpous moss *Brachythecium rutabulum* which exploits the herbaceous litter and shows a pronounced bimodal distribution with spring and autumnal maxima.

A rather different combination of phenologies is apparent in the study conducted by Tyler (1971) on a derelict sea-shore meadow (Figure 52) where the co-dominants *Juncus gerardii* and *Agrostis stolonifera* exhibit peaks of shoot biomass at quite different times in the spring and autumn, respectively.

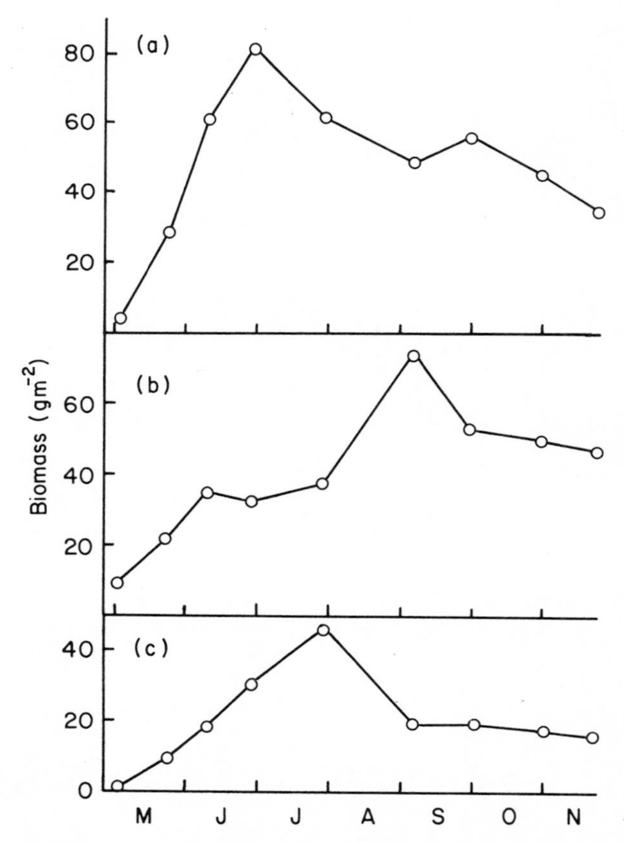

Figure 52 Seasonal changes in the shoot biomass of the main vegetation components in a derelict seashore meadow on the Baltic Coast: (a) *Juncus gerardii*, (b) *Agrostis stolonifera*, (c) *Eleocharis palustris* (Tyler, 1971). (Reproduced by permission of OIKOS.)

In species-rich vegetation such as that which occurs in ancient calcareous grasslands of the British Isles a more varied array of shoot phenologies may be observed (e.g. Figure 53). These range from the truncated vernal phenologies of geophytes, e.g. *Primula veris* and *Orchis mascula* (Figure 14), to certain of the evergreens such as *Carex flacca* (Figure 53) and *Koeleria cristata* in which it is often extremely difficult to detect any seasonal changes in shoot biomass. In

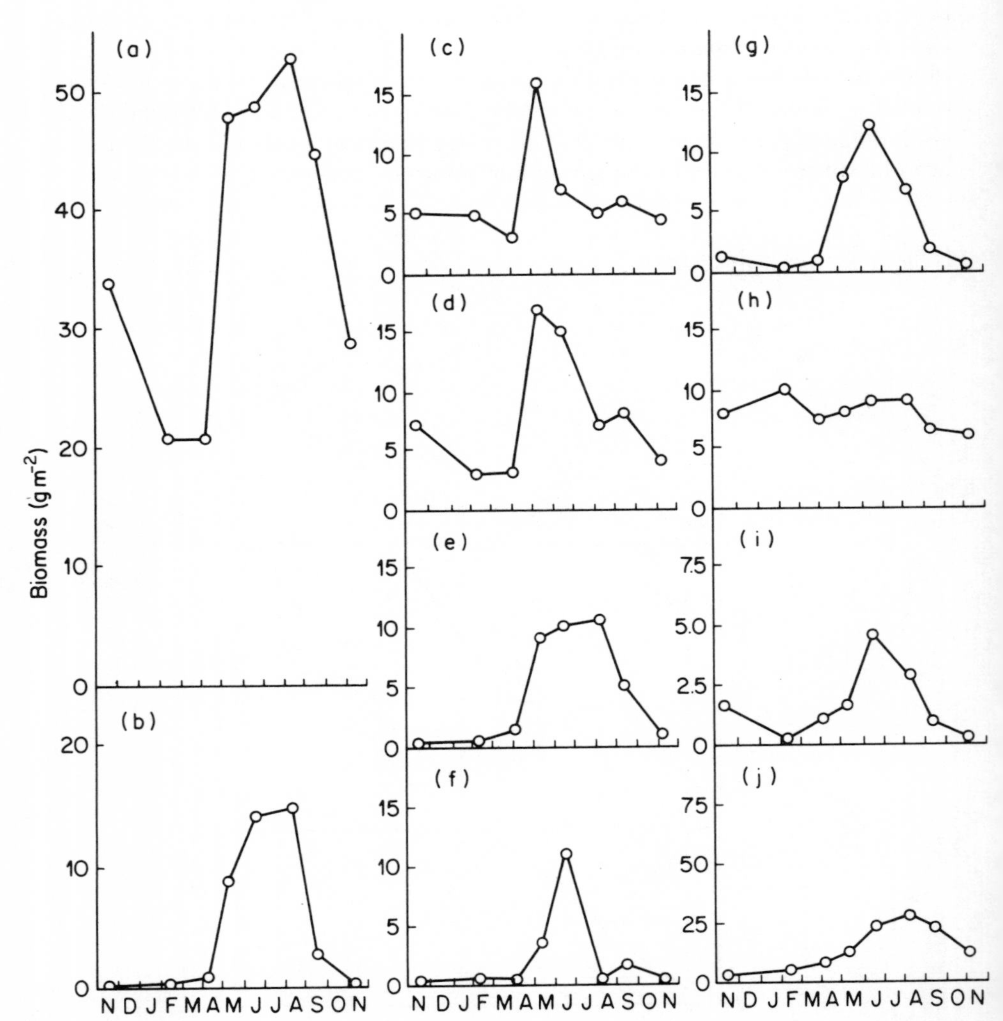

Figure 53 Seasonal changes in the shoot biomass of the main vegetation components in an area of derelict grassland subject to occasional burning on a very shallow calcareous soil in Northern England: (a) *Brachypodium pinnatum* (b) *Trifolium medium*, (c) *Carex caryophyllea*, (d) *Briza media*, (e) *Centaurea nigra*, (f) *Lotus corniculatus*, (g) *Leontodon hispidus*, (h) *Carex flacca*, (i) *Hieracium lachenalii*, (j) *Campanula rotundifolia* (Al-Mufti *et al.*, 1977). (Reproduced by permission of Blackwell Scientific Publications Ltd.)

some calcareous pastures there is a tendency for certain grasses (e.g. *Anthoxanthum odoratum*, *Festuca ovina*) to peak in growth earlier than forbs, such as *Poterium sanguisorba* (Figure 16), many of which possess long tap-roots (Figure 17) and exploit reserves of moisture during periods in which the grasses are subjected to desiccation. A similar distinction in phenology between grasses and forbs has been observed in a wide range of investigations in temperate and semi-arid regions (Getz, 1960; Golley, 1960, 1965; Menhinick, 1967; Barret, 1968; Precsenyi, 1969; Shure, 1971; Mellinger and McNaughton, 1975).

A review of the environmental factors and the physiological mechanisms which control shoot phenology is beyond the scope of this book. It is clear, however, that differences in response to seasonal variation in daylength and temperature are particularly important as determinants of the timing of growth, flowering, and vegetative reproduction. Where co-existence involves plants which grow in quite different seasons of the year, widely contrasted physiologies may be brought together in the same habitat. An example of this phenomenon is provided in Figure 54 from which it is apparent that the tall herb *Urtica dioica* and the moss *Brachythecium rutabulum* which are characteristically associated in the field (Plates 29 and 30) have markedly different temperature optima for growth.

So far the relationship of seasonal variation in environment to the phenomenon of co-existence in herbaceous vegetation has been considered exclusively in relation to established plants. However, there are many circumstances in which

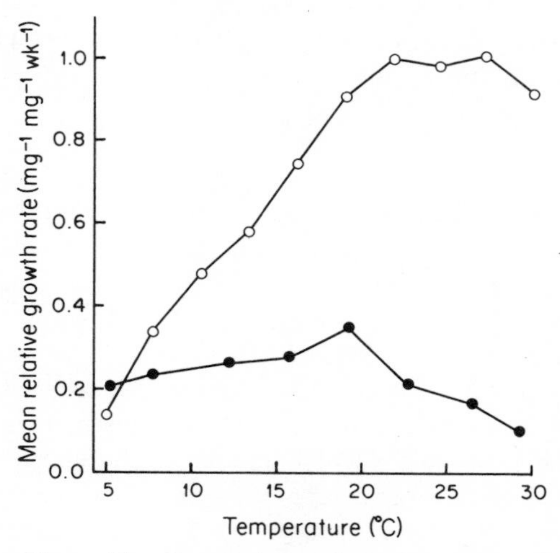

Figure 54 Comparison of the response to temperature in the relative growth-rates of the perennial herb *Urtica dioica* and the bryophyte *Brachythecium rutabulum*. Measurements on both species were conducted at a light intensity of 25 W m^{-2} (Al-Mufti and Furness, unpublished).

the presence of a variety of different species within the same small area of vegetation can be explained satisfactorily only by reference to the regenerative strategies of the species concerned. In Chapter 3, it was concluded that most forms of vegetative regeneration and seedling establishment depend upon colonization of vegetation gaps, and from the evidence which was reviewed it was apparent that, even between species which occupy similar habitats, there are major differences in the timing of gap exploitation.

In vegetation of high species density such as that occurring in unfertilized calcareous grasslands of the British Isles (Plate 31), gaps arise throughout the year from a wide range of phenomena. These include solifluction and frost-heaving during the winter and drought, rabbit-scraping, and the building of ant-hills during the summer. It is usually quite evident that the identities of the species which can most effectively colonize a gap change with season. The colonists of bare ground created in limestone pastures during the summer include the small winter annuals (page 64) and many of the grasses such as *Festuca ovina*, *Helictotrichon pratense*, and *Koeleria cristata* which produce large populations of seedlings in the early autumn. In the same habitat, colonization of gaps arising during the winter is usually delayed until the appearance in the following spring of populations of quite different species, e.g. *Linum catharticum*, *Pimpinella saxifraga*, and *Viola riviniana*.

A wide range of mechanisms account for the interspecific differences in germination times observed in species-rich vegetation. These include differences in time of seed release, in length of after-ripening period, in chilling requirement, and in response to temperature. An example of the diversity of responses to temperature which may be observed in seed populations from one type of plant community is illustrated in Figure 55.

Short-term variation

Although phenologies associated with season are of widespread importance, less conspicuous temporal niches operating on both shorter or longer time scales must be taken into account. Particularly in the British climate, daily, even hourly, fluctuations in radiation, temperature, and water potential of the atmosphere are likely to result in a constant shifting in the identity of the constituent species and genotypes for which conditions most nearly approximate to the optimum for photosynthesis and/or growth. Equally deserving of attention are the temporal niches which arise from short-term changes in vegetation structure resulting from grazing or mowing. Field observations and comparative studies (Milton, 1940; Mahmoud, 1973) have shown, for example, that there are profound differences between species and varieties of grasses with respect to their response to defoliation. From experiments such as that illustrated in Plate 32 there is evidence that the main reaction of *Lolium perenne* var. S23 to clipping consists of a rapid and almost vertical re-growth of the damaged leaves whereas, under the same treatment, *Agrostis tenuis* responds by producing a large number of small tillers and leaves, many of which are not projected into

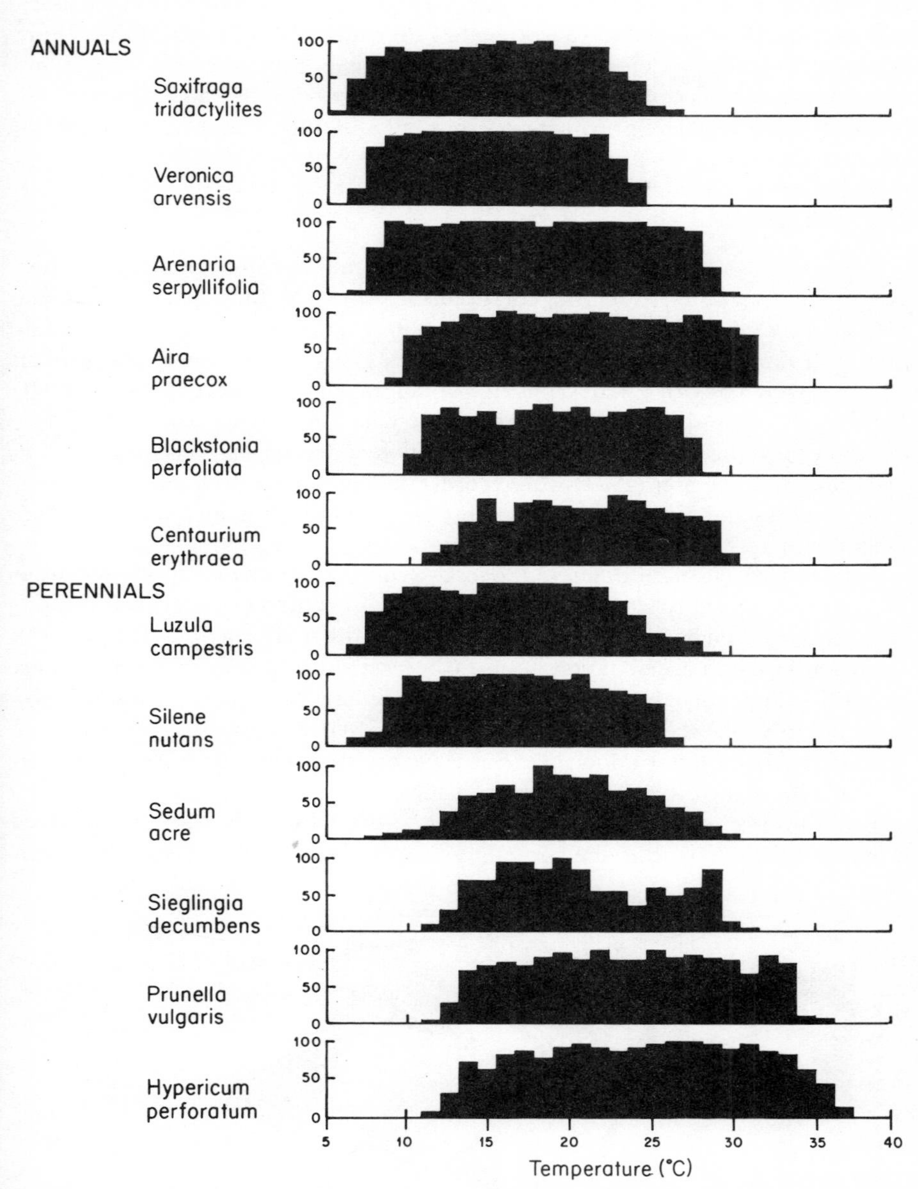

Figure 55 Comparison of the response to constant temperatures in the light in seeds collected from populations of herbaceous plants of common occurrence in calcareous grassland in Northern England. The figures tabulated refer to the final percentage germination at 30 temperatures over the range 5–40°C (light intensity 38.0 W m^{-2}, daylength 15 h).

the clipped stratum but instead form a compact low sward which includes stolons capable of invading bare patches and infiltrating areas of short turf occupied by other species. From these results it is apparent that conditions allowing co-existence of species such as *Lolium perenne* var. S23 and *Agrostis tenuis* are likely to arise where grazing is intermittent and fluctuations in the height of the turf alternately favour erect and prostrate growth-forms.

Long-term variation

Floristic diversity in herbaceous vegetation may be dependent also upon year-to-year variation in habitat conditions. Cyclical fluctuations in the ratio of *Lolium perenne* to *Trifolium repens* associated with changes in nitrogen status are a familiar example known to grassland ecologists (e.g. Leith, 1960) and similar effects correlated with fluctuations in rainfall have been reported (Watt, 1960, Hopkins 1978). An interesting study of the effect of year-to-year variation in habitat conditions upon the species composition of herbaceous vegetation is that of Müller and Foerster (1974) who recorded changes in the frequency of species in sown grasslands subject to varying intensities of flooding and silt deposition (Figure 56).

In order to explain many of the long-term fluctuations in the composition of a plant community it is necessary to identify the differences between component species which cause them to respond in different ways to year-to-year variation in environmental conditions. Contrasts in response to the conditions prevailing in particular years may arise from differences between species in the susceptibility of established plants to factors such as drought, predation, or pathogens. Even more important, however, are disparities in response caused by differences in regenerative characteristics.

Some insight into the diversity of regenerative capacities and requirements which may reside within one community can be gained from Figure 57 which

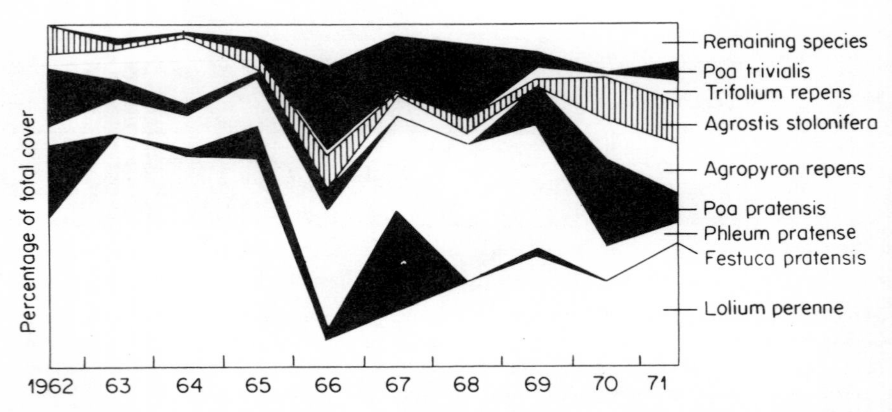

Figure 56 Changes in the species composition of a sown pasture subject to varying intensities of flooding and silt deposition in West Germany (Müller and Foerster, 1974). (Reproduced by permission of Verlag Paul Parey)

refers to two contrasted types of vegetation in Northern England. In both, it is apparent that within a small area of superficially homogeneous vegetation the component species exhibit a wide range of regenerative strategies, and it may be safely predicted that the identities of the species regenerating most successfully will change from year to year in accordance with fluctuations in climate and in factors such as the timing, distribution, form, and severity of vegetation disturbance. One source of variety in regenerative response which is conspicuous in Figure 57 is related to differences in capacity to develop persistent seed banks. In both of the plant communities for which data are presented, large and persistent seed banks are apparent in certain species, a number of which are relatively rare components of the established vegetation. It is reasonable to expect that these plants will become more abundant following years in which disturbance of the vegetation and the soil is exceptionally widespread or severe. In contrast to the plants with persistent seed banks, there are, in both communities, species, e.g. *Arrhenatherum elatius, Chamaenerion angustifolium, Festuca rubra*, in which the seeds remain dormant and/or viable for only a short period of time. The regenerative success of these plants in any year will depend crucially upon (1) the output of seeds and (2) the ability of the resulting offspring to exploit the particular opportunities for regeneration which characterize the year concerned.

Where species-rich and predominantly perennial herbaceous vegetation, e.g. limestone pasture, is subjected to year-to-year fluctuations in the intensity of climatic and biotic disturbance we may anticipate that the associated changes in species composition will be most apparent in the more ruderal component of the vegetation. Following the suggestion of Grubb (1976) we may suppose that when disturbance is restricted to the appearance of small gaps, tiny annuals and short-lived perennials such *Linum catharticum* will predominate. However, when the effect of severe drought, fire, or biotic disturbance is to cause the death of established perennials and to open large gaps it is not unusual for the resulting relaxation of stress to result in eruptions of competitive ruderals such as *Medicago lupulina* (Plate 33).

In certain forms of herbaceous vegetation of low productivity and high species density, such as the ancient limestone pastures of damp north-facing slopes in Northern Britain (Figure 9, Plate 31), the majority of the component species are long-lived perennials in which the investment in reproduction is small and intermittent. In these circumstances it is clear that the chances of consistently successful regeneration on the part of any particular species are extremely low and the mechanism maintaining diversity can be compared to an interminable game of roulette in which a large number of players place occasional and exceedingly prudent bets.

CO-EXISTENCE IN WOODY VEGETATION

Although there have been relatively few studies of co-existence between woody species there is reason to suspect that the mechanisms which control the

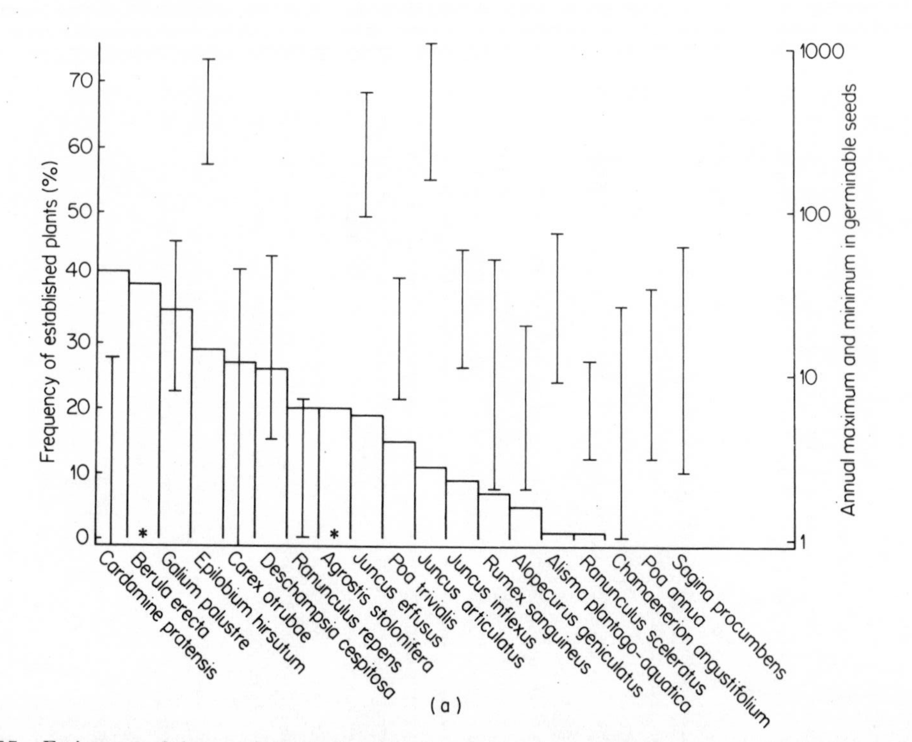

Figure 57 Estimates of the amplitude of seasonal variation in density of germinable seeds recovered from the top 3 cm of the soil profile (including litter) at two sites in Northern England: (a) a small eutrophic

of the vertical lines indicate the maximum and minimum numbers of seeds detected by a standardized procedure of sampling and laboratory germination applied at regular intervals throughout the year. The histograms describe the frequency of occurrence of established plants of the species present at the sites (Thompson and Grime, 1978). An asterisk means that no seeds were detected.

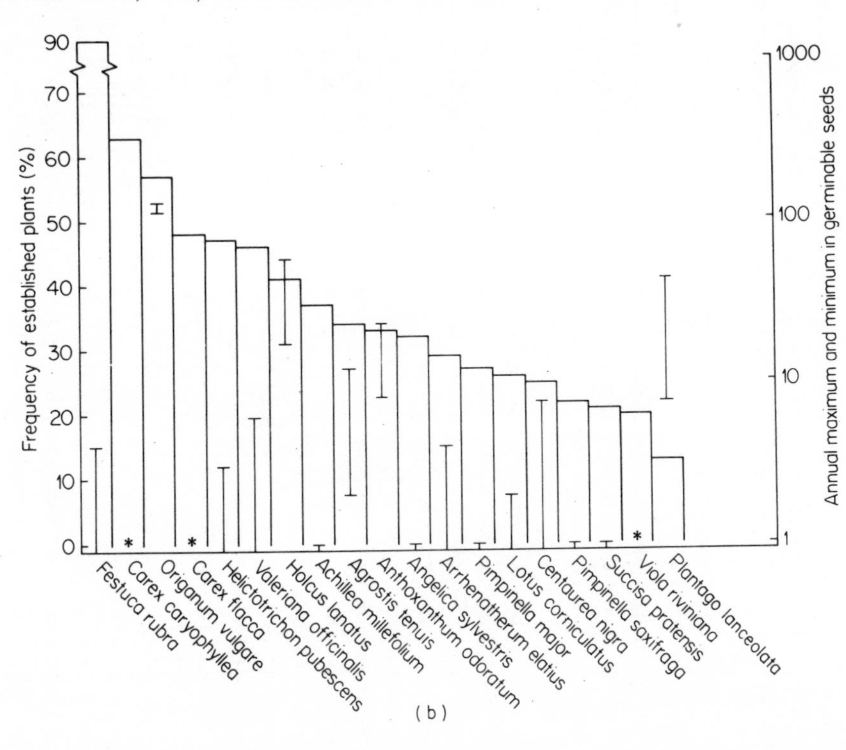

density of trees and shrubs occurring within areas of one hectare or less resemble in many respects those which operate on a smaller scale in herbaceous vegetation.

In temperate regions the highest densities of woody species occur in areas of open woodland and scrub on calcareous soils. In vegetation of this type in Southern Britain it is not unusual to find a mixture of shrubs including *Acer campestre, Corylus avellana, Euonymus europaeus, Ligustrum vulgare, Prunus spinosa, Thelycrania sanguineus,* and *Viburnum opulus*. On unproductive terrain such as steep, south-facing slopes with a discontinuous soil cover, diverse assemblages of these shrubs are often a persistent feature of the landscape. In these circumstances diversity is maintained, in part, by the low productivity of the habitat which restricts shrubs and trees to a size below that required to exercise major effects of dominance. In addition, it seems likely that spatial heterogeneity arising from variation in factors such as the depth, stability, and mineral nutrient status of the soil encourage the regeneration and persistence of a diversity of species.

In Britain and throughout the temperate zone a marked contrast with the species-rich calcareous scrub is provided by the various types of woodland which develop in areas where the soils are sufficiently stable and fertile to support the growth of large trees such as *Fagus sylvatica, Quercus petraea, Q. robur,* and *Acer pseudoplatanus*. Some of these species are capable of dominating extensive areas of woodland to the virtual exclusion of other trees and shrubs.

As in the case of herbaceous vegetation, therefore, we may conclude that high species density among trees and shrubs is promoted by moderately severe intensities of environmental stress. Evidence of this effect is not restricted to studies of temperate woodland: ecologists investigating the dynamics of tropical forests have concluded that here also there is an inverse relationship between productivity and species density. A particularly convincing example of this phenomenon is apparent in the work of Holdridge *et al.* (1971) from which it is clear that high densities of species of established trees are correlated with low mineral nutrient (especially phosphorus) status in forty samples of aboriginal tropical forest examined in Costa Rica.

The parallels between the mechanisms controlling species density in herbaceous and woody vegetation can be extended further by recognizing that high densities of trees and shrubs are not confined to unproductive habitats. In coppiced woodlands and hedgerows of the British Isles the effect of management is to so reduce the morphology of the potentially-dominant trees that vegetation types are produced in which 'debilitated' specimens of trees such as oak and beech may be found co-existing with a wide variety of small shrubs.

Differences in regenerative strategy also play a most important part in the vegetation processes which maintain species diversity in natural forests. It has been known for some time (Jones, 1945; Watt, 1947) that, in temperate deciduous woodland, co-existence by a variety of trees may be related to a continuous cyclic pattern of regeneration whereby openings in the canopy arising from the senescence and death of trees of one species tend to be colonized by seedlings or vegetative sprouts of another. In the Northern oakwoods of the British Isles, for example, gaps arising from the death of oak (*Quercus petraea*) frequently allow a temporary phase of recolonization by the wind-dispersed seeds of birch (*Betula pubescens*), which is itself then replaced by oak. Evidence of similar alternations of species has been obtained for both European forests (Nagel, 1950; Schaeffer and Moreau, 1958) and for mixed deciduous woodland in North America (Auclair and Cottam, 1971; Fox, 1977).

When attention is turned to the role of regenerative strategies in forests of high species density in temperate and tropical regions, very close similarities may be observed with the mechanisms of co-existence described for species-rich calcareous grassland (pages 176–7). For tropical forests there is an extensive literature documenting the importance of temporary gaps in providing opportunities for regeneration (Kramer, 1933; Richards, 1952; van Steenis, 1958; Schulz, 1960; Knight, 1975). In mature tropical forests where gaps arise mainly as a result of wind-fall (Jones, 1956; Cousens, 1965; Ricklefs, 1977) the most important forms of regeneration are those involving vegetative sprouts, banks of seedlings originating from fruits dispersed by animals, and, more rarely, wind-dispersed seeds. As in the case of the ancient limestone pastures discussed on page 177, many of the species produce seeds intermittently over a long life-history and rates of successful regeneration are exceedingly low.

Two additional features of regeneration in tropical forests provide strong parallels with phenomena occurring in herbaceous vegetation and are therefore relevant to a general model of the mechanism controlling species density. The first is based upon the observation (Blum, 1968; Gómez-Pompa, 1967; Webb, Tracey, and Williams, 1972) that loss of species density in severely disturbed tropical forests is often associated with widespread invasion by trees regenerating from buried seed banks (cf. the development of buried seed banks in disturbed areas of herbaceous vegetation and temperate forest (pages 89–90)). The second relates to the presence in certain types of mature tropical forest of trees with competitive characteristics. These exceedingly tall, deciduous, fast-growing 'nomads' (van Steenis, 1958; Whitmore, 1975; Knight, 1975) occur as scattered individuals which appear to depend for successful regeneration upon dispersal of wind-dispersed seeds into large openings in the canopy created by windfalls. It would appear that in rainforest it is only in such large gaps that the release from shade and mineral nutrient stress is sustained long enough to allow the establishment of competitive trees. In small gaps establishment is precluded by the rapid rate of canopy closure and the short duration of the pulse of mineral nutrients released by decomposition of fallen timber.

A close parallel may be drawn between the nomads of tropical rainforest and competitive-ruderals such as *Medicago lupulina* (Plate 33) which often occur as large isolated individuals in temperate grasslands dominated by slow-growing perennial herbs. As noted on page 177, scrutiny of the distribution of *M. lupulina* in unproductive grassland confirms that, as in the case of the rainforest nomads, this species is confined to sites in which local damage has created exceptionally large gaps in the established vegetation.

CONCLUSIONS

A general model

In Figure 58 an attempt has been made to summarize the relationships between five processes which influence species density in vegetation. As in the model already considered on page 163 (Figure 46), three main contingencies are

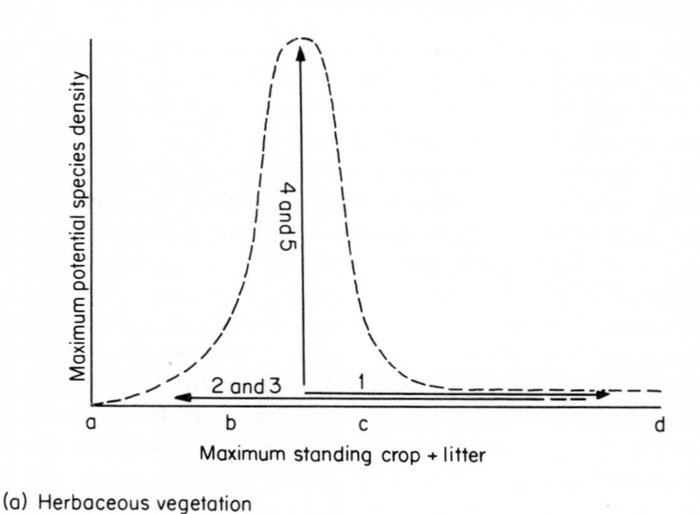

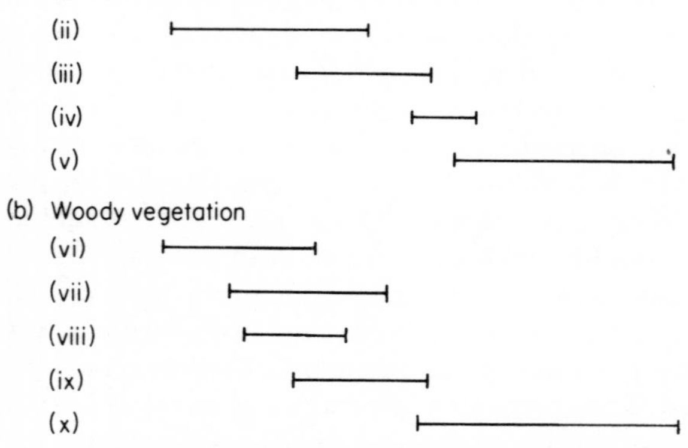

Figure 58 Model summarizing the impact of five processes upon the species density in vegetation. Key to processes: 1, Dominance; 2, stress; 3, disturbance; 4, niche-differention; 5, ingress of suitable species or genotypes. The horizontal lines describe the range of contingencies encompassed by a number of familiar herbaceous and woody vegetation types. Key to vegetation types: (i) paths, (ii) grazed rock outcrops with discontinuous soil cover, (iii) infertile pastures, (iv) fertilized pastures, (v) derelict fertile pastures, (vi) arctic scrub, (vii) temperate scrub, (viii) temperate hedgerows, (ix) species-rich tropical rainforest, (x) mature temperate or tropical forest on fertile soil.

recognized, each corresponding to a different section of the baseline of the figure. The section a–b includes the species-poor vegetation found in environments subjected to extreme conditions of stress and/or disturbance. The corridor containing vegetation types with the potential for relatively high species density occupies the intermediate part of the range in biomass (b–c), whilst the section c–d corresponds to vegetation in which species densities are suppressed by dominance.

Two processes can be identified which determine the species densities attained in particular corridor environments. The first of these has been examined in some detail earlier in this chapter and consists of the degree of spatial and temporal variation and the resulting opportunities for complementary forms of exploitation and regeneration within the environment. The second process, which will be considered under the next heading, is the availability and rate of ingress of potential constituent species from the surrounding landscape.

In the lower part of Figure 58 an attempt has been made to illustrate the relationship of the model to patterns of variation in species density which can be observed in the field. Horizontal lines have been drawn to indicate those parts of the model which correspond to the range of conditions in particular habitats or vegetation types. It must be emphasized that although the model may be used to interpret variation in species density within either herb or tree layers, there will be differences of detail in the way in which the mechanisms controlling species density operate on herbs and trees. In particular, it is clear that in the tree layer the thresholds (a,b,c) at which various phenomena come into play along the baseline of Figure 58 will correspond to very much higher values in biomass. Moreover, for the reasons discussed on page 165, we may expect that there will be quantitative differences between tropical and temperate vegetation types.

In passing, it is interesting to note that studies of species density and community structure in intertidal algal communities (Dayton, 1970; Paine, 1969, 1974) and benthic algae (Lewis, 1914; MacFarlane and Bell, 1933; Hehre and Mathieson, 1970; Sears and Wilce, 1975; Dayton and Hessler, 1972) have recognized many of the phenomena included in the general model presented here. There is also evidence which suggests that many of the same principles may be applied to the control of species density in colonial animals such as corals (Branham *et al.*, 1971; Dana *et al.*, 1972; Goreau *et al.*, 1972), shell-fish (Connell, 1961, 1972), and even some noncolonial organisms such as coral reef fish (Sale, 1977).

'Reservoir effects' upon species density

So far it has been convenient to analyse the control of species density mainly in terms of the structure of vegetation and the interactions of the component plants both with each other and with various features of the environment. Already, however (page 165), it has been necessary to refer to the role of an additional factor which may exercise a major limitation on species density. This is the reservoir of suitable species in each geographical area. Particularly in extensively-disturbed landscapes such as that of Lowland Britain, certain habitats such as woodland or unproductive calcareous pasture or marshland have been reduced by agricultural development to small isolated fragments. Where these fragments are of recent origin it is frequently observed that their potential for high species density remains unfulfilled and we may suppose that this is, in part, because of the slow rate of ingress of suitable plants from the adjacent countryside. In these circumstances, low species densities appear to be determined by the low dispersal efficiencies of the seeds of many plants and also by the depauperate state of the surrounding flora.

A rather different reservoir effect appears to be involved in the consistent differences in species density which are observed in temperate regions between adjacent and structurally-similar vegetation types on calcareous and acidic soils (e.g. Plate 34). From data such as those illustrated in Figure 44, for example, it is apparent that in Britain, regardless of habitat, species densities in herbaceous vegetation are consistently low where the surface soil pH is less than 4.0. A more specific example of the reduction in species density associated with increasing soil acidity is presented in Figure 59 which is based upon survey data from an area of Northern England in which there is an intimate mosaic in which acidic and calcareous grasslands experience comparable effects of climate and land-use. From this figure it is evident that whereas high species densities are attained over the pH range 4.0–8.0, there is an abrupt decline on more acidic soils. When variation in species density across the soil pH range is compared with the total number of species recorded from all the samples examined in each soil pH category it is apparent that the fall in species density in individual samples is correlated with a progressive reduction in the size of the species reservoir associated with the transition from calcicoles to calcifuges. Reference to the *Flora of the British Isles* (Clapham, Tutin, and Warburg, 1962) and the *Atlas of the British Flora* (Perring and Walters, 1962) confirms that the greater abundance of calcicoles over calcifuges in Britain applies not only to grassland herbs but to other ecological groups such as trees, shrubs, and ruderal plants. This suggests that the maximum species densities which can be attained in any particular vegetation type will show a predictable relationship to soil pH. In Figure 60 an attempt has been made to summarize the relationships between maximum species density, maximum standing crop + litter and soil pH in herbaceous vegetation of the British Isles.

We may conclude, therefore, that at latitudes such as that of the British Isles, the higher species densities supported by calcareous soils are related, in part, to the fact that in such regions calcicolous vegetation draws upon a reservoir of species which is considerably larger than that of the calcifuges. The most likely explanation for the greater abundance of calcicoles is that these plants have evolved mainly at lower latitudes where the effect of low precipitation : evaporation ratios is to maintain a high base-status in the soils. This is to suggest that the evolution of calcicoly has occurred in semi-arid environments where, as suggested by Margalef (1968), Stebbins (1952, 1972), Stebbins and Major

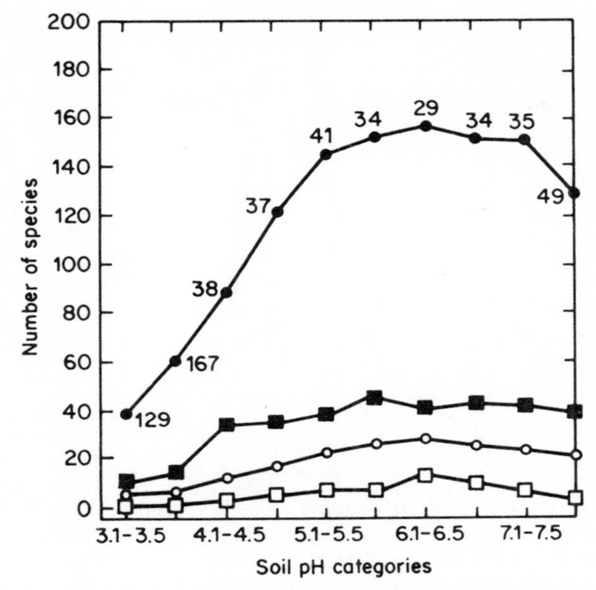

Figure 59 Mean (ō), maximum (▪), and minimum (▫) species density and total number of species represented (•) in categories of surface soil pH encountered in 593 one square metre quadrats located at random in unmanaged grasslands distributed within an area of 2400 km² in Northern England. The values inserted on the topmost curve refer to the number of quadrats falling in each half-unit soil pH category (Grime, 1973c). (Reproduced with permission from *J. Envir. Man.*, 1, 151–167. Copyright 1973 by Academic Press Inc. (London) Ltd.)

(1965), and Bartholomew, Eaton, and Raven (1973), the effects of climatic fluctuation are to maintain a greater degree of vegetation disturbance and environmental heterogeneity, with higher rates of turnover in plant populations, all of which are conducive to relatively rapid rates of speciation and diversification of floras.

Latitudinal gradients in the size of the reservoir of angiosperms extend from the polar regions to the equator (Figure 61), and this fact has important implications for the control of species density in a variety of vegetation types. In forests, for example, it is well known that whereas in temperate regions the mean density of trees is less than 10 species/Ha, values exceeding 100 species/Ha are frequently encountered in tropical rainforest (Longman and Janik, 1974). It seems reasonable to suggest, therefore, that the higher densities of tree species in tropical forests are related to the larger reservoir of trees present at low latitudes and this, in turn, may be attributed to the long and uninterrupted period over which forest speciation and co-adaptation (Dobzhansky, 1950; Janzen, 1970; Gilpin, 1975) has been able to take place in equatorial habitats.

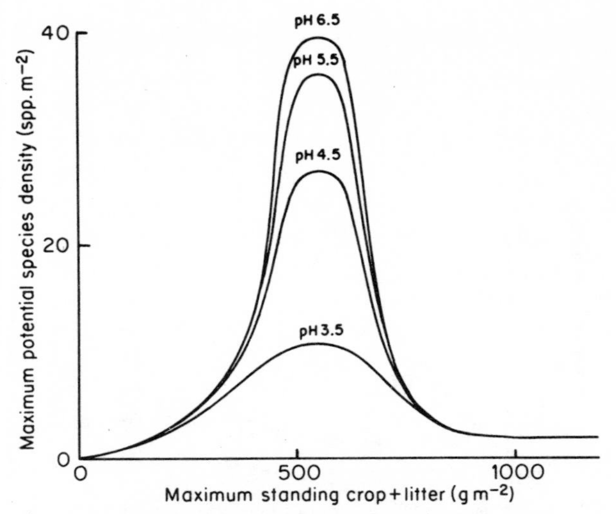

Figure 60 Diagram summarizing the relationship between surface soil pH, seasonal maximum in standing crop + litter, and maximum potential species density in herbaceous vegetation of the British Isles.

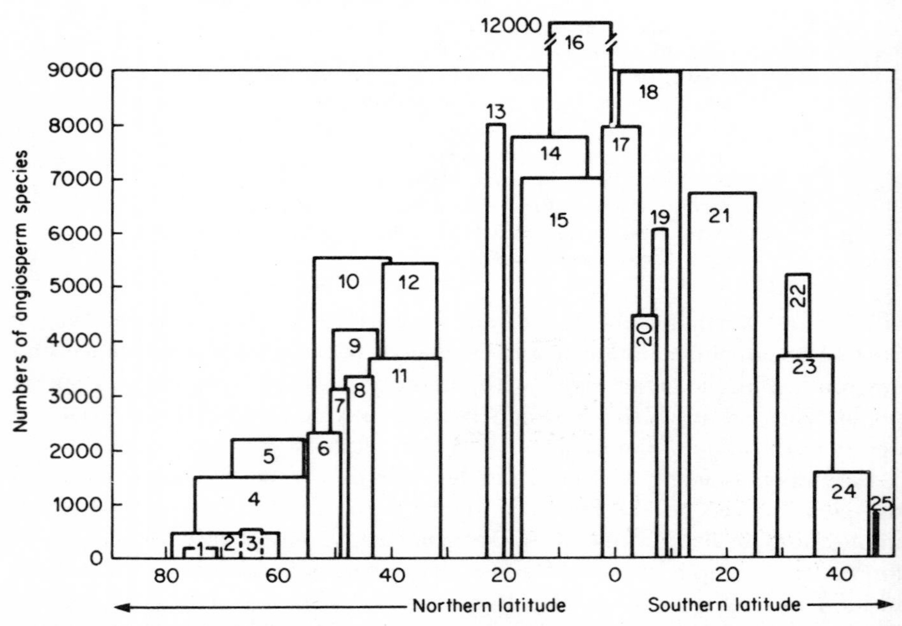

Figure 61 Diagram illustrating the latitudinal gradient in species richness of angiosperms. 1, Novaya Zemlya; 2, Greenland; 3, Iceland; 4, Yakut ASSR; 5, Sweden; 6, Poland; 7, Czechoslovakia; 8, Roumania; 9, France; 10, Kasakhstan; 11, Japan; 12, California; 13, Cuba; 14, Philippine; 15, River-basins of Niger and Volta; 16, Venezuela; 17, Gabon; 18, New Guinea; 19, Java; 20, Central Tanzania; 21, Madagascar; 22, Cape Peninsula; 23, New South Wales; 24, New Zealand; 25, Prince Edward Island. (Rejmánek, 1976) (Reproduced by permission of Plenum Publishing Corporation.)

The contribution of epiphytes to species-rich vegetation

There are a number of vegetation types in which the exceptionally high species density depends to a considerable extent upon the presence of a rich epiphytic component. Reference has been made already to the epiphytic habit of certain of the bryophytes in herbaceous vegetation (pages 66 and 169), and in temperate regions many of the most ancient woodlands are characterized by the diverse assemblages of mosses and lichens on the trunks and branches of the older trees (Rose, 1974). However, it is in tropical rainforests that epiphytes achieve their greatest biomass and taxonomic variety.

In the tropical rainforest environment, the contrasted types of micro-habitat which occur at different heights above ground are exploited by a wide range of lichens, bryophytes, pteridophytes, and angiosperms. The abundance of the epiphytic flora may be related in particular to the conditions of moisture and mineral nutrient supply within the rainforest canopy. Because of the frequent rainfall and high humidity, germination, establishment and survival can take place in the virtual absence of a rooting medium (Plate 35), and in some extreme climates epiphytes colonize not only trunks and branches but also the surfaces of actively-photosynthetic leaves (Pessin, 1922). Moreover, it seems reasonable to expect that under such rainfall conditions there will be a constant leaching of solutes from the tree canopy; there is a need to investigate the mechanisms whereby rainforest epiphytes are adapted to exploit this source of mineral nutrients.

Recently, attention has been drawn to an additional mechanism whereby epiphytes may influence the floristic diversity of tropical rainforest. It has been suggested (Strong, 1977) that in some rainforests the burden of epiphytes may be sufficient to induce high rates of windfalls thus creating the gaps exploited by seedlings and vegetative sprouts of many of the trees.

Control of species density by vegetation management

Management of vegetation, whether for the purposes of agriculture, amenity, landscape restoration, or nature conservation, inevitably brings the practical ecologist into contact with many of the phenomena discussed in this chapter. As an illustration of the relevance of these theoretical concepts to vegetation management two examples in quite different fields of applied ecology will now be considered briefly

Maintenance of monocultures in agricultural systems

Since much time, energy, and money is expended in the attempt to maintain monocultures of crops and forage plants it is of more than academic interest to attempt to establish where particular agricultural systems lie in relation to the various contingencies summarized in Figure 58. With respect to pastures and meadows in Britain, a valuable clue is provided by the evidence (page 125) of the increasing dominance and reductions in species density when productivity is stimulated by heavy dressings of mineral fertilizers. When this effect is portrayed diagrammatically (Figure 62), one is prompted to consider whether by further modifications of farming regimes it would be possible to move even closer to the conditions favouring monoculture. As we have seen (page 125–129), herbaceous monocultures or 'near-monocultures' occur in non-agricultural situations in Britain. However, since these are composed of plants which owe their dominant status to relatively undisturbed conditions it may be unrealistic to expect to be able to create vegetation of this type in cropped systems. It remains to be determined, therefore, whether stable perennial monocultures are attainable and consistent with the objectives of farm management. Nevertheless, it is established that regimes which allow the development of a large standing crop and minimize the frequency of cropping will encourage the trend towards monoculture.

Management of vegetation subject to trampling

A recurrent problem in managing parkland and areas of natural landscape where herbaceous vegetation is subject to trampling is to estimate the carrying capacity of different habitats and to predict the response of vegetation to increasing intensities of wear. It is interesting to note that in the research reported in this field there is apparently conflicting evidence concerning the impact of trampling in grassland. Some investigators (e.g. Bayfield, 1973; Liddle and Greig-Smith, 1976) have recorded a fall in species density with increasing intensity of trampling whilst others (e.g. Westhoff, 1967; van der Maarel, 1971) have detected a change in the reverse direction. However, as

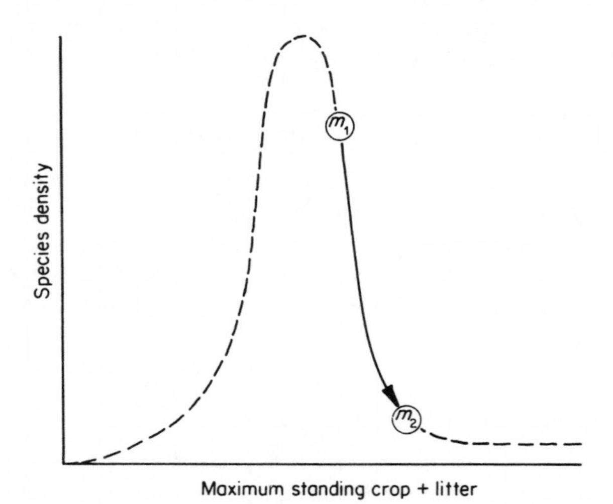

Figure 62 Diagram representing the alteration in species density associated with the vegetation change ($m_1 \rightarrow m_2$) resulting from the application of a heavy dressing of mineral fertilizer to a species-rich meadow.

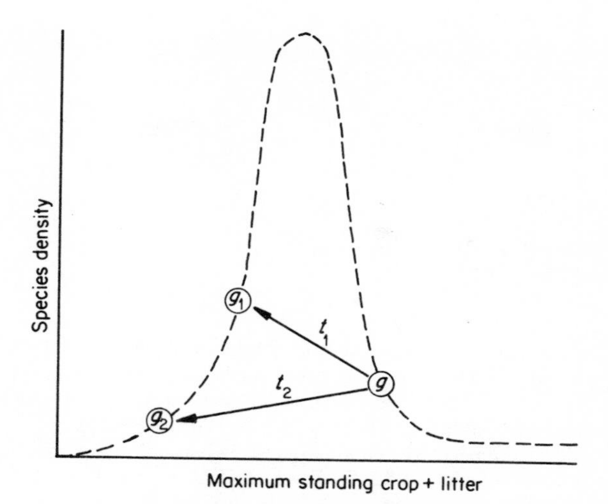

Figure 63 Diagram representing the alteration of species density which is associated with the changes in vegetation ($g \rightarrow g_1$, $g \rightarrow g_2$) brought about when relatively undisturbed base-rich grassland is subjected to moderate (t_1) and severe (t_2) intensities of trampling.

pointed out by Liddle (1975), both of these types of response can be explained by reference to the 'hump-backed' model. This point is illustrated in Figure 63 which depicts the hypothetical responses of a productive and relatively undisturbed grassland community to two intensities of trampling. At the lower intensity of trampling the result is a reduction in the vigour of the dominant components of the turf allowing an ingress of species. The effect of more severe trampling, however, is to create an environment in which only a small number of specialized plants are able to survive.

References

Abdul-Wahab, A. S. and E. L. Rice (1967). Plant inhibition by Johnson grass and its possible significance in old field succession. *Bull. Torrey Bot. Club*, **94**, 986–997.

Abrahamson, W. G. and M. Gadgil (1973). Growth form and reproductive effort in goldenrods (*Solidago*, Compositae). *Amer. Natur.*, **107**, 651–661.

Allard, R. W. (1965). Genetic systems associated with colonizing ability in predominantly self-pollinated species. In H. G. Baker and G. L. Stebbins (Eds) *The Genetics of Colonizing Species*, Academic Press, New York, pp. 49–75.

Allee, W. C. (1938). *The social life of animals*, Norton, New York.

Al-Mashhadani, Y. D. (1979). Experimental investigations of competition and allelopathy in herbaceous plants. Ph.D. Thesis, University of Sheffield.

Al-Mufti, M. M. (1978). A quantitative and phenological study of the herbaceous vegetation in Totley Wood. Ph.D. Thesis, University of Sheffield.

Al-Mufti, M. M. and J. P. Grime (1979). The disruptive effect of deciduous tree canopies on woodland ground-floras. *New Phytol.*, **82** (in press).

Al-Mufti, M. M., C. L. Sydes, S. B. Furness, J. P. Grime and S. R. Band (1977). A quantitative analysis of shoot phenology and dominance in herbaceous vegetation. *J. Ecol.*, **65**, 759–791.

Ampofo, S. T., K. G. Moore, and P. H. Lovell (1976). The role of the cotyledons in four *Acer* species and in *Fagus sylvatica* during early seedling development, *New Phytol.*, **76**, 31–39.

Arno, S. F. and J. R. Habeck (1972). Ecology of Alpine Larch (*Larix lyalli*, Parl.) in the Pacific Northwest. *Ecol. Monogr.*, **42**, 417–450.

Ashton, D. H. (1958). The ecology of *Eucalyptus regnans* F. Muell: the species and its frost resistance. *Australian J. Bot.*, **6**, 154–176.

Ashton, D. H. and B. J. Macauley (1972). Winter leaf spot disease of seedlings of *Eucalytus regnans* and its relation to forest litter. *Trans. Brit. Mycol. Soc.*, **58**, 377–386.

Auclair, A. N. and G. Cottam (1971). Dynamics of black cherry (*Prunus serotina* Erhr.) in Southern Wisconsin oak forests. *Ecol. Monogr.*, **41**, 153–177.

Baker, H. G. (1965). Characteristics and modes of origin of weeds. In H. G. Baker and G. L. Stebbins (Eds) *The Genetics of Colonizing Species*, Academic Press, New York, pp. 147–168.

Baker, H. G. and G. L. Stebbins (Eds)(1965). *The Genetics of Colonizing Species*, Academic Press, New York.

Bakker, D. (1960). A comparative life-history of *Cirsium arvense* (L) Scop. and *Tussilago farfara* (L), the most troublesome weeds in the newly reclaimed polders of the former Zuiderzee. In J. L. Harper (Ed.), *The Biology of Weeds, Symp, Br. Ecol. Soc.*, **1**, 205–222.

Barrett, G. W. (1968). The effects of an acute insecticide stress on a semi-enclosed grassland ecosystem. *Ecology*, **49**, 1019–1035.

Bartholomew, B., L. C. Eaton, P. V. Raven (1973). *Clarkia rubicunda*: a model of plant evolution in semi-arid regions. *Evolution*, **27**, 505–517.

Barton, L. V. (1961). *Seed preservation and longevity*, Hill, London.

Batzli, G. O. and F. A. Pitelka (1970). Influence of meadow mouse populations on California grassland. *Ecology*, **51**, 1027–1035.
Bayfield, N. G. (1973). Use and deterioration of some Scottish hill paths. *J. appl. Ecol.*, **10**, 639–648.
Beadle, N. C. W. (1954). Soil phosphate and the delimitation of plant communities in eastern Australia, I. *Ecology*, **35**, 370–375.
Beadle, N. C. W. (1962). Soil phosphate and the delimitation of plant communities in eastern Australia, II. *Ecology*, **43**, 281–288.
Beatley, J. C. (1967). Survival of winter annuals in the northern Mojave Desert. *Ecology*, **48**, 745–750.
Bentley, P. J. (1966). Adaptations of amphibia to arid environments. *Science*, **152**, 619–623.
Berg, R. Y. (1954). Development and dispersal of the seed of *Pedicularis silvatica*. *Nytt. Mag. Bot.*, **2**, 1–58.
Bernard, J. M. and J. G. McDonald (1974). Primary production and life-history of *Carex lacustris*. *Can. J. Bot.*, **52**, 117–123.
Bernard, J. M. and B. A. Solsky (1977). Nutrient cycling in a *Carex lacustris* wetland. Can. J. Bot., **55**, 630–638.
Bernhard, F. (1970). Etude de la litière et de sa contribution au cycle des éléments mineraux en forêt ombrophile de Cote-d'Ivoire. *Oecol. Plant.*, **5**, 247–266.
Bertalanaffy, L. von (1950). The theory of open systems in physics and biology. *Science*, **111**, 23–29.
Bibbey, R. O. (1948). Physiological studies of weed seed germination. *Pl. Physiol. Lancaster*, **23**, 467–484.
Billings, W. D. and P. J. Godfrey (1968). Acclimation effects on metabolic rates of arctic and alpine *Oxyria* populations subjected to temperature stress (Abstr.). *Bull. Ecol. Soc. Am.*, **49**, 68–69.
Billings, W. D., P. J. Godfrey, B. F. Chabot, and D. P. Bourgue (1971). Metabolic acclimation to temperature in arctic and alpine ecotypes of *Oxyria digyna*. *Arctic Alpine Res.*, **3**, 277–289.
Billings, W. D. and H. A. Mooney (1968). The ecology of arctic and alpine plants. *Biol. Rev.*, **43**, 481–529.
Birrel, K. S. and A. C. S. Wright (1945). A serpentine soil in New Caledonia. *New Zeal. J. Sci. Technol.*, **27A**, 72–76.
Björkman, O. (1968a). Carboxydismutase activity in shade-adapted and sun-adapted species of higher plants. *Physiol. Plant.*, **21**, 1–10.
Björkman, O. (1968b). Further studies on differentiation of photosynthetic properties in sun and shade ecotypes of *Solidago virgaurea*. *Physiol. Plant.*, **21**, 84–99.
Björkman, O. and P. Holmgren (1963). Adaptability of the photosynthetic apparatus to light intensity in ecotypes from exposed and shaded habitats. *Physiol. Plant.*, **16**, 889–914.
Björkman, O. and P. Holmgren (1966). Adaptation to light intensity in plants native to shaded and exposed habitats. *Physiologia Pl.*, **19**, 854–859.
Black, J. M. (1958). Competition between plants of different initial seed sizes in swards of subterranean clover (*Trifolium subterraneum* L.) with particular reference to leaf area and the light micro-climate. *Australian J. Agr. Res.*, **9**, 299–312.
Blackman, G. E. and J. N. Black (1959). Physiological and ecological studies in the analysis of plant environment: XI. A further assessment of the influence of shading on the growth of different species in the vegetative phase. *Ann. Bot., N. S.*, **23**, 51–63.
Blackman, G. E. and A. J. Rutter (1948). Physiological and ecological studies in the analysis of plant environment: III. The interaction between light intensity and mineral nutrient supply in leaf development and in the net assimilation rate of the bluebell (*Scilla non-scripta*). *Ann. Bot., N. S.*, **12**, 1–6.

Blackman, G. E. and G. L. Wilson (1951a). Physiological and ecological studies in the analysis of plant environment: VI. The constancy for different species of a logarithmic relationship between net assimilation rate and light intensity and its ecological significance. *Ann. Bot., N. S.*, **15**, 63–94.

Blackman, G. E. and A. J. Wilson (1951b). Physiological and ecological studies in the analysis of plant environment: VII. An analysis of the differential effects of light intensity on the net assimilation rate, leaf–area ratio and relative growth rate of different species. *Ann. Bot., N. S.*., **15**, 373–408.

Bliss, L. C. (1971). Arctic and alpine life cycles. *Ann. Rev. Ecol. Syst.*, **2**, 405–438.

Blum, K. E. (1968). Contributions toward an understanding of vegetational development in the Pacific lowlands of Panama. Ph.D. Thesis, Florida State University, Tallahassee.

Böcher, T. W. (1949). Racial divergencies in *Prunella vulgaris* in relation to habitat and climate. *New Phytol.*, **48**, 285–314.

Böcher, T. W. (1961). Experimental and cytological studies in plant species: VI. *Dactylis glomerata* and *Anthoxanthum odoratum*. *Bot. Tidskr.*, **56**, 314–355.

Böcher, T. W. and K. Larsen (1958). Geographical distribution of initiation of flowering, growth habit, and other characters in *Holcus lanatus*. *Bot. Notiser.*, **3**, 289–300.

Bordeau, P. F. and M. L. Laverick (1958). Tolerance and photosynthetic adaptability to light intensity in white pine, red pine, hemlock, and ailanthus seedlings. *Forest Sci.*, **4**, 196–207.

Bostock, S. J. (1976). The life history strategies of selected perennial Compositae of disturbed ground. Ph.D. Thesis, University of Manchester.

Boysen-Jensen, P. (1929). Studier over Skovtracerres Forhold til Lyset. *Dansk. Skovforen. Tidssk.*, **14**, 5–31.

Boysen-Jensen, P. (1932). *Die Stoffproduktion der Pflanzen*, Jena.

Bradbury, I. K. and G. Hofstra (1976). The partitioning of net energy resources in two populations of *Solidago canadensis* during a single developmental cycle in southern Ontario. *Can. J. Bot.*, **54**, 2449–2456.

Bradshaw, A. D. (1959). Population differentiation in *Agrostis tenuis* Sibth: I. Morphological differentiation. *New Phytol.*, **58**, 208–227.

Bradshaw, A. D. (1977). Conservation problems in the future. In *Scientific Aspects of Nature Conservation in Great Britain, Proc. Roy. Soc. Lond. B.*, **197**, 77–96.

Bradshaw, A. D., M. J. Chadwick, D. Jowett, and R. W. Snaydon (1964). Experimental investigations into the mineral nutrition of several grass species: IV. Nitrogen level. *J. Ecol.*, **52**, 665–676.

Bragg, A. N. (1945). The spadefoot toads in Oklahoma with a summary of our knowledge of the group. *Amer. Natur.*, **79**, 52–72.

Branham, J. M., S. Reed, J. H. Bailey, and J. Caperon (1971). Coral-eating sea stars *Acanthaster planci* in Hawaii. *Science*, **172**, 1155–1157.

Bratton, S. P. (1976). Resource division in an understory herb community: responses to temporal and microtopographic gradients. *Amer. Natur.*, **110**, 679–693.

Bray, J. R. (1958). Notes toward an ecologic theory. *Ecology*, **39**, 770–776.

Brenchley, W. E. (1918). Buried weed seeds. *J. Agric. Sci.*, **9**, 1–31.

Brenchley, W. E. and K. Warington. (1930). The weed seed population of arable soil: I. Numerical estimation of viable seeds and observations on their natural dormancy. *J. Ecol.*, **18**, 235–272.

Brenchley, W. E. and K. Warington (1958). *The Park Grass Plots at Rothamsted 1856–1949*, Rothamsted Experimental Station, Harpenden.

Bronson, F. W. (1964). Agonistic behaviour in wood-chucks. *Anim. Behav.*, **12**, 270–278.

Brown, E. S. (1962). The African armyworm, *Spodoptera exempta* (Walker) (*Lepidoptera, Noctuidae*): a review of the literature, Commonwealth Institute of Entomology, London.

Brown, W. H. (1919). *The Vegetation of Philippine Mountains*, Bur. Sci., Manila.

Bunting, E. S. (1956). An Agronomic Study of *Papaver somniferum* L. Ph.D. Thesis, University of London.

Burgess, P. F. (1972). Studies on the regeneration of the hill forests of the Malay Peninsula. *Malay Forest*, **35**, 103–123.

Burns, G. P. (1923). Studies in tolerance of New England forest trees: IV. Minimum light requirement referred to a definite standard. *Bull. Vermont. Agr. Exp. Sta.*, **235**, 15–25.

Büttner, R. (1971). Untersuchungen zur Okologie und Physiologie des Gasstoffwechsels bei einigen Strauchflechten. *Flora (Jena)*, **160**, 72–99.

Callaghan, T. V. (1976). Growth and population dynamics of *Carex bigelowii* in an alpine environment. Strategies of growth and population dynamics of tundra plants, 3. *Oikos*, **27**, 402–413.

Calow, P. (1977). The joint effect of temperature and starvation on the metabolism of triclads. *Oikos*, **29**, 87–92.

Calow, P. and A. S. Woolhead (1977). The relationship between ration, reproductive effort, and age-specific mortality in the evolution of life-history strategies — some observations on freshwater triclads. *J. Anim. Ecol.*, **46**, 765–781.

Cates, R. G. and G. H. Orians (1975). Successional status and the palatability of plants to generalized herbivores. *Ecology*, **56**, 410–418.

Cavers, P. B. and J. L. Harper (1966). Germination polymorphism in *Rumex crispus* and *Rumex obtusifolius*. *J. Ecol.*, **54**, 307–382.

Champness, S. S. and K. Morris (1948). Populations of buried viable seeds in relation to contrasting pasture and soil types. *J. Ecol.*, **36**, 149–173.

Chang, C. F., A. Suzuki, S. Kumai, and S. Tamura (1969). Chemical studies on 'clover sickness': II. Biological functions of isoflavenoids and their related compounds. *Agric. biol. Chem.*, **33**, 398–408.

Chapin, F. S. and A. Bloom (1976). Phosphate absorption: adaptation of tundra graminoids to a low temperature, low phosphorus environment. *Oikos*, **26**, 111–121.

Chew, R. M. and B. B. Butterworth (1964). Ecology of rodents in Indian Cove (Mojave Desert), Joshua Tree National Monument, California. *J. Mammal.*, **45**, 203–225.

Chew, R. M. and A. E. Chew (1970). Energy relationships of the mammals of a desert shrub (*Larrea tridentata*) community. *Ecol. Mongr.*, **40**, 1–21.

Chippendale, H. G. and W. E. J. Milton (1934). On the viable seeds present in the soil beneath pastures. *J. Ecol.*, **22**, 508–531.

Clapham, A. R., T. G. Tutin, and E. F. Warburg (1962). *Flora of the British Isles*, 2nd ed., Cambridge University Press, London.

Clarkson, D. T. (1967). Phosphorus supply and growth rate in species of *Agrostis* L. *J. Ecol.*, **55**, 111–118.

Clausen, J., D. D. Keck, and W. M. Hiesey (1940). Experimental studies on the nature of species: I. Effect of varied environments on western North American plants. *Carnegie Inst. Washington Pub. 520*.

Clements, F. E. (1916). *Plant succession. An analysis of the development of vegetation*, Carnegie Inst., Washington.

Cody, M. (1966). A general theory of clutch size. *Evolution*, **20**, 174–184.

Cole, L. C. (1954). The population consequences of life history phenomena. *Quart. Rev. Biol.*, **29**, 103–137.

Connell, J. H. (1961). Effect of competition, predation by *Thais lappillus* and other factors on natural populations of the barnacle *Balanus balanoides*. *Ecol. Monogr.*, **31**, 61–104.

Connell, J. H. (1972). Community interactions on marine intertidal shores. *Ann. Rev. Ecol. Syst.*, **3**, 169–192.

Cook, R. E. (1976). Photoperiod and the determination of potential seed number in *Chenopodium rubrum* L. *Ann. Bot.*, **40**, 1085–1099.

Cook, S. C. A., C. Lefèbvre, and T. McNeilly (1972). Competition between metal tolerant and normal plant populations on normal soil. *Evolution*, **26**, 366–372.

Cottam, W. P., J. M. Tucker, and R. Drobnick (1959). Some clues to Great Basin postpluvial climates provided by oak distributions. *Ecology*, **40**, 361–377.

Cousens, J. E. (1965). Some reflections on the nature of Malayan lowland rain forest. *Malay. For.*, **28**, 122–128.

Crisp, M. D. and R. T. Lange (1976). Age structure, distribution and survival under grazing of the arid-zone shrub *Acacia burkittii*. *Oikos*, **27**, 86–92.

Cumming, B. G. (1969). *Chenopodium rubrum* L. and related species. In L. T. Evans (Ed.), *The Induction of Flowering*, Cornell University Press, Ithaca, N.Y.

Currey, D. R. (1965). An ancient bristlecone pine stand in Eastern Nevada. *Ecology*, **46**, 564–566.

Curtis, J. T. and G. Cottam (1950). Antibiotic and autotoxic effects in priarie sunflower. *Bull. Torrey Bot. Club*, **77**, 187–191.

Dana, T. F., W. A. Newman, and E. W. Fager (1972). *Acanthaster* aggregations: Interpreted as primarily responses to natural phenomena. *Pac. Sci.*, **26**, 355–372.

Daniel, C. P. and R. B. Platt (1968). Direct and indirect effects of short term ionizing radiation on old field succession. *Ecol. Monogr.*, **38**, 1–29.

Darwin, C. (1859). *The origin of species by means of natural selection or the preservation of favoured races in the struggle for life*, Murray, London.

Davis, R. F. (1928). The toxic principle of *Juglans nigra* as identified with synthetic juglone and its toxic effects on tomato and alfalfa plants. *Amer. J. Bot.*, **15**, 620.

Davison, A. W. (1964). Some factors affecting seedling establishment on calcareous soils. Ph.D. Thesis, University of Sheffield.

Davison, A. W. (1977). The ecology of *Hordeum murinum* L: III. Some effects of adverse climate. *J. Ecol.*, **65**, 523–530.

Dawson, J. H. (1970). *Time and duration of weed infestations in relation to weed–crop competition*, Proceedings of the 23rd Southern Weed Conference, pp. 13–35.

Dayton, P. K. (1970). Competition, disturbance and community organization: The provision and subsequent utilization of space in a rocky inter-tidal community. *Ecol. Monogr.*, **41**, 351–389.

Dayton, P. K. and R. R. Hessler (1972). Role of biological disturbance in maintaining diversity in the deep sea. *Deep-sea Res.*, **19**, 199–208.

Del Moral, R. and C. H. Muller (1969). Fog drip: a mechanism of toxin transport from *Eucalyptus globulus*. *Bull. Torrey Bot. Club.*, **96**, 467–475.

DeWit, C. T. (1960). On competition. *Versl. Landbouwkd. Onderz.*, **66**, 1–82.

DeWit, C. T. (1961). Space relationships within populations of one or more species. In F. L. Milthorpe (Ed.), *Mechanisms in Biological Competition*, Cambridge University Press, London, pp. 314–329.

Dixon, S. (1892). The effects of settlement and pastoral occupation in Australia upon the indigenous vegetation. *Trans. R. Soc. S. Aust.*, **15**, 195–206.

Dobzhansky, T. (1950). Evolution in the tropics. *Amer. Sci.*, **38**, 209–221.

Donald, C. M. (1958). The interaction of competition for light and for nutrients. *Australian J. Agr. Res.*, **9**, 421–432.

Drew, M. C., L. R. Saker, and T. W. Ashley (1973). Nutrient supply and the growth of the seminal root system. *J. exp. Bot.*, **24**, 1189–1202.

Drury, W. H. and I. C. T. Nisbet (1973). Succession. *J. Arnold Arboretum Harvard Univ.*, **54**, 331–368.

Duffey, E., E. G. Morris, J. Sheail, L. K. Ward, D. A. Wells, and T. C. E. Wells (1974). *Grassland Ecology and Wildlife Management*, Chapman and Hall, London.

Egunjobi, J. K. (1974a). Dry matter, nitrogen and mineral element distribution in an unburnt savanna during the year. *Oecol. Plant*, **9**, 1–10.

Egunjobi, J. K. (1974b). Litter fall and mineralization in a teak (*Tectona grandis*) stand. *Oikos*, **25**, 222–226.

Ellenberg, H. (1963). *Vegetation Mitteleuropas mit den Alpen*, Eugen Ulmer, Stuttgart.

Ellenberg, H. and D. Mueller-Dombois (1967a). Tentative physiognomic-ecological classification of plant formations of the earth. *Ber. geobot. Inst. ETH, Stiftg. Rübel*,

Zürich, **37**, 21–55. (Republished 1969 as UNESCO report SC/WS/269 in slightly modified form, entitled 'A framework for a classification of world vegetation,' Paris, 26 pp. Finalized UNESCO publication 1973. International classification and mapping of vegetation. Ecology and Conservation series No. 6, 93 pp and chart with map symbols.)

Ellenberg, H. and D. Mueller-Dombois (1967b). A key to Raunkiaer plant life forms with revised subdivisions. *Ber. geobot. Inst. ETH, Stiftg. Rubel, Zurich*, **37**, 56–73.

Ellenberg, H. and D. Mueller-Dombois (1974). *Aims and methods of vegetation ecology*, Wiley, New York.

Elton, C. S. and R. S. Miller (1954). The ecological survey of animal communities: with a practical system of classifying habitats by structural characteristics. *J. Ecol.*, **42**, 460–496.

Evenari, M. (1965). Physiology and seed dormancy, after-ripening and germination. *Proc. Int. Seed Test Assoc.*, **30**, 49–71.

Farrar, J. F. (1976a). Ecological physiology of the lichen *Hypogymnia physodes*: II. Effects of wetting and drying cycles and the concept of physiological buffering. *New Phytol.*, **77**, 105–113.

Farrar, J. F. (1976b). Ecological physiology of the lichen *Hypogymnia physodes*: III. The importance of the rewetting phase. *New Phytol.*, **77**, 115–125.

Farrar, J. F. (1976c). The uptake and metabolism of phosphate by the lichen *Hypogymnia physodes*. *New Phytol.*, **77**, 127–134.

Feeny, P. P. (1968). Effects of oak leaf tannins on larval growth of the winter moth *Operophtera brumata*. *J. Insect Physiol.*, **14**, 805–817.

Feeny, P. P. (1969). Inhibitory effect of oak leaf tannins on the hydrolysis of proteins by trypsin. *Phytochemistry*, **8**, 2119–2126.

Feeny, P. P. (1970). Seasonal changes in oak leaf tannins and nutrients as a cause of spring feeding by winter moth caterpillars. *Ecology*, **5**, 565–581.

Feeny, P. P. (1975). Biochemical coevolution between plants and their insect herbivores. In L. E. Gilbert and P. H. Raven (Eds.), *Coevolution of Plants and Animals*, University of Texas Press, pp. 3–19.

Fenner, M. J. (1978). A comparison of the competitive abilities of ruderals and closed-turf species. *J. Ecol.* (submitted).

Ferguson, C. W. (1968). Bristlecone Pine: science and esthetics. *Science*, **159**, 839–846.

Fisher, H. J., L. F. Myers, and J. D. Williams (1974). Nutrient responses of an indigenous *Poa* Tussock and *Lolium perenne* L. grown separately and together in pot culture. *Aust. J. Agric. Res.*, **25**, 863–874.

Fisher, R. A. (1930). *The genetical theory of natural selection*, Clarendon, Oxford.

Ford, E. D. and P. J. Newbould (1977). The biomass and production of ground vegetation and its relation to tree cover through a deciduous woodland cycle. *J. Ecol.*, **65**, 201–212.

Foulds, W. (1978). Response to soil water potential in three leguminous species: I. Growth, reproduction and mortality. *New Phytol.*, **79** (in press).

Fox, J. F. (1977). Alternation and coexistence of tree species. *Amer. Nat.*, **111**, 69–89.

French, N. R. (1967). Life spans of *Dipodomys* and *Perognathus* in the desert. *J. Mammal.*, **48**, 537–548.

French, N. R., B. G. Maza, and A. P. Aschwanden (1966). Periodicity of desert rodent activity. *Science*, **154**, 1194–1195.

French, N. R., B. G. Maza, H. O. Hill, A. P. Aschwanden, and H. W. Kaaz (1974). A population study of irradiated desert rodents. *Ecol. Monogr.*, **44**, 45–72.

Frith, H. J. (1973). Discussion: role of environment in reproduction as a source of 'predictive' information. In D. S. Farner (Ed.), *Breeding Biology of Birds*, National Academy of Sciences, Washington, D.C., pp. 147–154.

Furness, S. B. (1979). Ecological effects of temperature in bryophytes and herbaceous plants. Ph.D. Thesis, University of Sheffield.

Furness, S. B., S. J. van de Dijk, and J. P. Grime (1979). An investigation of the

exploitation of herbaceous litter by the pleurocarpous moss, *Brachythecium rutabulum*. *J. Ecol.*, **66**, (in press).
Gadgil, M. and O. T. Solbrig (1972). The concept of r- and K-selection: evidence from wild flowers and some theoretical considerations. *Amer. Natur.*, **106**, 14–31.
Gardner, G. (1977). The reproductive capacity of *Fraxinus excelsior* on the Derbyshire limestone. *J. Ecol.*, **65**, 107–118.
Gardner, R. A. and K. E. Bradshaw (1954). Characteristics and vegetative relationships of some podzolic soils near the coast of northern California. *Soil Sci. Soc. Am. Proc.*, **18**, 320–325.
Gates, D. M. (1968). Transpiration and leaf temperature. *Ann. Rev. Plant Physiol.*, **19**, 211–238.
Getz, L. L. (1960). Standing crop of herbaceous vegetation is southern Michigan. *Ecology*, **41**, 393–395.
Gilpin, M. E. (1975). Limit cycles in competition communities. *Amer. Natur.*, **109**, 51–60.
Godwin, H. (1956). *The history of the British flora*, Cambridge University Press, London.
Golley, F. B. (1960). Energy dynamics of a food chain of an old-field community. *Ecol. Monogr.*, **30**, 187–206.
Golley, F. B. (1965). Structure and function of an old-field broomsedge community. *Ecol. Monogr.*, **35**, 113–137.
Gómez-Pompa, A. (1967). Some problems of tropical forest plant ecology. *J. Arnold Arbor.*, **48**, 105–121.
Goreau, N. I. and C. M. Yonge (1971). Reef corals: autotrophs or heterotrophs? *Biol. Bull. Marine Biol. Lab. Woods Hole*, **141**, 247–260.
Goreau, T. F., J. C. Lang, E. A. Graham, and P. D. Goreau (1972). Structure and ecology of the Saipan reefs in relation to predation by *Acanthaster planci* (L). *Bull. Marine Sci.*, **22**, 113–152.
Gorski, T. (1975). Germination of seeds in the shadow of plants. *Physiol. Plant.*, **34**, 342–346.
Grace, J. and H. W. Woolhouse (1970). A physiological and mathematical study of the growth and productivity of a *Calluna-Sphagnum* community: I. Net photosynthesis of *Calluna vulgaris* (L.) Hall. *J. Appl. Ecol.*, **7**, 363–381.
Grant, P. R. (1970). Experimental studies of competitive interactions in a two-species system: II. The behavior of *Microtus, Peromyscus* and *Clethrionomys* species. *Anim. Behav.*, **18**, 411–426.
Grieve, B. J. (1956). Studies in the water relations of plants: I. Transpiration of western Australian (Swan Plain) sclerophylls. *J. Roy. Soc. Western Australia*, **40**, 15–20.
Grime, J. P. (1963). An ecological investigation at a junction between two plant communities in Coombsdale on the Derbyshire Limestone. *J. Ecol.*, **51**, 391–402.
Grime, J. P. (1965). Comparative experiments as a key to the ecology of flowering plants. *Ecology*, **45**, 513–515.
Grime, J. P. (1966). Shade avoidance and tolerance in flowering plants. In R. Bainbridge, G. C. Evans, and O. Rackman (Eds.), *Light as an Ecological Factor*, Blackwell, Oxford, pp. 281–301.
Grime, J. P. (1972). The creative approach to nature conservation. In F. J. Ebling and G. W. Heath (Eds.), *The Future of Man*, Academic Press, London, pp. 47–54.
Grime, J. P. (1973a). Competitive exclusion in herbaceous vegetation. *Nature*, **242**, 344–347.
Grime, J. P. (1973b). Competition and diversity in herbaceous vegetation — a reply. *Nature*, **244**, 310–311.
Grime, J. P. (1973c). Control of species density in herbaceous vegetation. *J. Envir. Man.*, **1**, 151–167.
Grime, J. P. (1974). Vegetation classification by reference to strategies. *Nature*, **250**, 26–31.
Grime, J. P. (1977). Evidence for the existence of three primary strategies in plants and its relevance to ecological and evolutionary theory. *Amer. Natur.*, **111**, 1169–1194.

Grime, J. P. (1978). Interpretation of small-scale patterns in the distribution of plant species in space and time. In J. W. Woldendorp (Ed.), *A synthesis of demographic and experimental approaches to the functioning of plants*, Wageningen, 101–124.

Grime, J. P. and G. M. Blythe (1968). An investigation of the relationships between snails and vegetation at the Winnats Pass. *J. Ecol.*, **57**, 45–66.

Grime, J. P. and A. V. Curtis (1976). The interaction of drought and mineral nutrient stress in calcareous grassland. *J. Ecol.*, **64**, 976–998.

Grime, J. P. and B. C. Jarvis (1975). Shade avoidance and shade tolerance in flowering plants: II. Effects of light on the germination of species of contrasted ecology. In R. Bainbridge, G. C. Evans, and O. Rackman (Eds.), *Light as an Ecological Factor*, Blackwell, Oxford, pp. 525–532.

Grime, J. P. and D. W. Jeffrey (1965). Seedling establishment in vertical gradients of sunlight. *J. Ecol.*, **53**, 621–642.

Grime, J. P. and R. Hunt (1975). Relative growth rate: its range and adaptive significance in a local flora. *J. Ecol.*, **63**, 393–422.

Grime, J. P. and P. S. Lloyd (1973). *An Ecological Atlas of Grassland Plants*, Arnold, London.

Grime, J. P., S. F. MacPherson-Stewart, and R. S. Dearman (1968). An investigation of leaf palatability using the snail *Cepaea nemoralis* L. *J. Ecol.*, **56**, 405–420.

Grouzis, M., A. Berger, and G. Heim (1976). Polymorphisme et germination des graines chez trois espécès anuelles du genre *Salicornia*. *Oecol. Plant*, **11**, 41–52.

Grubb, P. J. (1976). A theoretical background to the conservation of ecologically distinct groups of annuals and biennials in the chalk grassland ecosystem. *Biol. Conserv.*, **10**, 53–76.

Grubb, P. J. (1977). The maintenance of species-richness in plant communities: the importance of the regeneration niche. *Biol. Rev.*, **52**, 107–145.

Grubb, P. J., H. E. Green, and R. C. J. Merrifield (1969). The ecology of chalk heath; its relevance to the calcicole–calcifuge and soil acidification problems. *J. Ecol.*, **57**, 175–212.

Grümmer, G. and H. Beyer (1960). The influence exerted by species of *Camelina* on flax by means of toxic substances. In J. L. Harper (Ed.), *The Biology of Weeds*, Blackwell, Oxford, pp. 153–157.

Gutterman, Y. (1974). The influence of the photoperiodic regime and red–far-red light treatments of *Portulaca oleracea* L. plants on the germinability of their seeds. *Oecologia (Berl.)*, **17**, 27–38.

Haag, R. W. (1974). Nutrient limitations to plant production in two tundra communities. *Can. J. Bot.*, **52**, 103–116.

Hackett, C. (1965). Ecological aspects of the nutrition of *Deschampsia flexuosa* (L.)Trin: II. The effects of Al, Ca, Fe, K, Mn, N, P, and pH on the growth of seedlings and established plants. *J. Ecol.*, **53**, 315–333.

Hadley, E. B. and L. C. Bliss (1964). Energy relationships of alpine plants on Mt. Washington, New Hampshire. *Ecol. Monogr.*, **34**, 331–357.

Hanes, T. L. (1971). Succession after fire in the chaparrel of southern California. *Ecol. Monogr.*, **41**, 27–52.

Hansen, K. (1976). Ecological studies in Danish heath vegetation. *Dansk. Bot. Arkiv.*, **41**, 1–118.

Harley, J. L. (1969). *The biology of Mycorrhiza*, 2nd ed. Hill, London.

Harley, J. L. (1970). Mycorrhiza and nutrient uptake in forest trees. In L. Luckwell and C. Cutting (Eds.), *Physiology of tree crops*, Academic Press, New York, pp. 163–178.

Harley, J. L. (1971). Fungi in ecosystems. *J. Ecol.*, **59**, 653–686.

Harper, J. L. (1957). Biological flora of the British Isles: *Ranunculus acris* L., *Ranunculus repens* L., and *Ranunculus bulbosus* L. *J. Ecol.*, **45**, 289–342.

Harper, J. L. (1961). Approaches to the study of plant competition. In F. L. Milthorpe (Ed.), *Mechanisms in Biological Competition*, *Symp. Soc. exp. Biol., 15.* Cambridge University Press, pp. 1–39.

Harper, J. L. (1977). *The population biology of plants*, Academic Press, London.
Harper, J. L. and J. White (1974). The demography of plants. *A. Rev. Ecol. Syst.*, **5**, 419–463.
Harrington, J. F. and R. C. Thompson (1952). Effect of variety and area of production on subsequent germination of lettuce seed at high temperatures. *Proc. Am. Soc. Hort. Sci.*, **59**, 445–450.
Harris, P. (1960). Production of pine resin and its effect on survival of *Ryhacionia beolina. Can. J. Zool.*, **38**, 121–130.
Hehre, E. J. and A. C. Mathieson (1970). Investigations of New England marine algae: III. Composition, seasonal occurrence and reproductive periodicity of the marine Rhodophyceae in New Hampshire. *Rhodora*, **72**, 194–239.
Hellmers, H. (1964). An evaluation of the photosynthetic efficiency of forests. *Quart. Rev. Biol.*, **39**, 249–257.
Hewitt, L. J. (1952). A biological approach to the problems of soil acidity. *Trans. Int. Soc. Soil Sci. Jt. Meet. Dublin*, **1**, 107–118.
Hicklenton, P. R. and W. C. Oechel (1976). Physiological aspects of the ecology of *Dicranum fuscescens* in the subarctic: I. Acclimation and acclimation potential of CO_2 exchange in relation to habitat, light and temperature. *Can. J. Bot.*, **54**, 1104–1119.
Hickman, J. C. (1975). Environmental unpredictability and plastic energy allocation strategies in the annual *Polygonum cascadense* (Polygonaceae). *J. Ecol.*, **63**, 689–701.
Higgs, D. E. B. and D. B. James (1969). Comparative studies on the biology of upland grasses: I. Rate of dry matter production and its control in four grass species. *J. Ecol.*, **57**, 553–563.
Hodgson, G. L. and G. E. Blackman (1956). An analysis of the influence of plant density on the growth of *Vicia faba. J. Exp. Bot.*, **7**, 146–165.
Hoffman, G. R. (1966). Ecological studies of *Funaria hygrometrica* (L.) Hebw. in Eastern Washington and Northern Idaho. *Ecol. Monogr.*, **36**, 157–180.
Holdridge, L. R., W. C. Grenke, W. H. Hatheway, T. Liang, and J. A. Tosi, Jr. (1971). *Forest environments in Tropical Life Zones: a pilot study*, Pergamon Press, Oxford.
Hopkins, B. (1966). Vegetation of the Olokemeji Forest Reserve, Nigeria: IV. The litter and soil with special reference to their seasonal changes. *J. Ecol.*, **54**, 687–703.
Hopkins, B. (1978). The effects of the 1976 drought on chalk grassland in Sussex, England. *Biol. Conserv.*, **14**, 1–12.
Horn, H. S. (1971). *The Adaptive Geometry of Trees*, Princeton University Press.
Horn, H. S. (1974). The ecology of secondary succession. *Annual. Rev. Ecol. Syst.*, **5**, 25–37.
Hosakawa, T., N. Odani, and H. Tagawa (1964). Causality of the distribution of corticolous species in forests with special reference to the physiological approach. *Bryologist*, **67**, 396–411.
Howe, H. F. (1977). Bird activity and seed dispersal of a tropical wet forest tree. *Ecology*, **58**, 539–550.
Hughes, M. K. (1975). Ground vegetation net production in a Danish beech wood. *Oecologia (Berl.)*, **18**, 251–258.
Hughes, R. D. (1970). The seasonal distribution of bushfly (*Musca vetustissime* Walker) in south-east Australia. *J. Anim. Ecol.*, **39**, 691–706.
Hutchinson, G. E. (1951). Copepodology for the ornithologist. *Ecology*, **32**, 571–577.
Hutchinson, G. E. (1959). Homage to Santa Rosalia or why are there so many kinds of animals?, *Amer. Natur.*, **93**, 145–159.
Hutchinson, T. C. (1967). Comparative studies of the ability of species to withstand prolonged periods of darkness. *J. Ecol.*, **55**, 291–299.
Ingestad, T. (1973). Mineral nutrient requirements of *Vaccinium vitis-idaea* and *V. myrtillus. Physiol. Plant.*, **29**, 239–246.
Janzen, D. H. (1970). Herbivores and the number of tree species in tropical forests. *Amer. Natur.*, **104**, 501–528.

Janzen, D. H. (1971). Seed predation by animals. *Ann. Rev. Ecol. Syst.*, **2**, 265–292.

Janzen, D. H. (1973). Community structure of secondary compounds in plants. In T. Swain (Ed.), *Chemistry in Evolution and Systematics*, Russak, New York, pp. 529–538.

Janzen, D. H. (1976). The depression of reptile biomass by large herbivores. *Amer. Natur.*, **110**, 371–400.

Jarvis, P. G. and M. S. Jarvis (1964). Growth rates of woody plants. *Physiologia Pl.*, **17**, 654–666.

Jeffrey, D. W. (1967). Phosphate nutrition of Australian heath plants: I. The importance of proteoid roots in *Banksia* (Proteaceae). *Aust. J. Bot.*, **15**, 403–411.

Jeffrey, D. W. (1971). The experimental alteration of a *Kobresia*-rich sward in Upper Teesdale. In E. Duffey and A. S. Watt (Eds.), *The Scientific Management of Animal and Plant Communities for Conservation*, Blackwell, Oxford, pp. 78–89.

Jeffries, R. L. and A. J. Willis (1964). Studies on the calcicole–calcifuge habit: II. The influence of calcium on the growth and establishment of four species in soil and sand cultures. *J. Ecol.*, **52**, 691–707.

Jenny, H., S. P. Gessel, and F. T. Bingham (1949). Comparative study of decomposition rates of organic matter in temperate and tropical regimes. *Soil Sci.*, **68**, 419–432.

Jones, E. W. (1945). Structure and reproduction of the virgin forest of the north temperate zone. *New Phytol.*, **44**, 130–148.

Jones, E. W. (1956). Ecological studies on the rainforest of southern Nigeria: IV. The plateau forest of the Okumu Forest Reserve. *J. Ecol.*, **44**, 83–117.

Jones, R. K. (1974). A study of the phosphorus responses of a wide range of accessions from the genus *Styolanthes*. *Aust. J. Agric. Res.*, **25**, 847–62.

Jordan, C. F. and J. R. Kline (1972). Mineral cycling: some basic concepts and their application in a tropical rain forest. *Ann. Rev. Ecol. Syst.*, **3**, 33–50.

Jowett, D. (1964). Population studies on lead-tolerant *Agrostis tenuis*. *Evolution*, **18**, 70–80.

Junttila, O. (1973). Seed and embryo germination in *Syringa vulgaris* and *S. reflexa* as affected by temperature during seed development. *Physiologia Pl.*, **29**, 264–268.

Karssen, C. M. (1968). The light-promoted germination of the seeds of *Chenopodium album*; effects of (RS)-abscisic acid. *Acta. bot. neerl.*, **17**, 293–308.

Katznelson, J. (1972). Studies in clover soil sickness: I. The phenomenon of soil sickness in berseem and Persian clover. *Pl. Soil*, **36**, 379–393.

Keast, A. (1959). Australian birds: their zoogeography and adaptation to an arid continent. In A. Keast, R. L. Crocker, and C. S. Christian (Eds.), *Biogeography and ecology in Australia*, Junk, The Hague, pp. 89–114.

Keay, R. W. J. (1957). Wind-dispersed species in a Nigerian forest. *J. Ecol.*, **45**, 471–478.

Keever, C. (1973). Distribution of major forest species in southeastern Pennsylvania. *Ecol. Monogr.*, **43**, 303–327.

Kershaw, K. A. (1977a), Physiological–environmental interactions in lichens: II. The pattern of net photosynthetic acclimation in *Peltigera canina* (L.) Willd var *praetextata* (Floerke in Somm.) Hue, and *P. polydactyla* (Neck.) Hoffm. *New Phytol.*, **79**, 377–390.

Kershaw, K. A. (1977b). Physiological–environmental interactions in lichens: III. The rate of net photosynthetic acclimation in *Peltigera canina* (L.) Willd. var *praetextata* (Floerke in Somm.) Hue, and *P. polydactyla* (Neck.) Hoffm. *New Phytol.*, **79**, 391–402.

King. T. J. (1975). Inhibition of seed germination under leaf canopies in *Arenaria serpyllifolia, Veronica arvensis* and *Cerastium holosteoides*. *New Phytol.*, **75**, 87–90.

King, T. J. (1976). The viable seed content of ant-hill and pasture soil. *New Phytol.*, **77**, 143–147.

King, T. J. (1977a). The plant ecology of ant-hills in calcareous grasslands: I. Patterns of species in relation to ant-hills in southern England. *J. Ecol.*, **65**, 235–256.

King, T. J. (1977b). The plant ecology of ant-hills in calcareous grasslands: II. Succession on the mounds. *J. Ecol.*, **65**, 257–278.

King, T. J. (1977c). The plant ecology of ant-hills in calcareous grasslands: III. Factors affecting the population sizes of selected species. *J. Ecol.*, **65**, 279–316.

Kingsbury, R. W., A. Radlow, P. J. Mudie, J. Rutherford, and R. Radlow (1976). Salt stress responses in *Lasthenia glabrata*, a winter annual composite endemic to saline soils. *Can. J. Bot.*, **54**, 1377–1385.
Knight, D. H. (1975). A phytosociological analysis of species-rich tropical forest on Barro Colorado Island, Panama. *Ecol. Monogr.*, **45**, 259–284.
Koford, C. B. (1958). Prairie dogs, white faces and blue grama. *Wildl. Monogr.*, **3**, 78 pp.
Koller, D. (1962). Preconditioning of germination in lettuce at time of fruit ripening. *Am. J. Bot.*, **49**, 841–844.
Koller, D. (1969). The physiology of dormancy and survival of plants in desert environments. In H. W. Woolhouse (Ed.), *Dormancy and Survival*, Symposium of the Society for Experimental Biology, **23**, 449–469.
Kramer, F. (1933). De natuurlijke verjonging in het Goenoeng-Gedehcomplex. *Tectona*, **26**, 156–185.
Kropac, Z. (1966). Estimation of weed seeds in arable soil. *Pedobiologia*, **6**, 105–128.
Kruckeberg, A. R. (1954). The ecology of serpentine soils: III. Plant species in relation to serpentine soils. *Ecology*, **35**, 267–274.
Kubiček, F. and J. Brechtl (1970). Production and phenology of the herb layer in an oak-hornbeam forest. *Biologia*, **25**, 651–666.
Lack, D. (1947). The significance of clutch size. Parts I and II. *Ibis*, **89**, 302–352.
Lack, D. (1948). The significance of clutch size. Part III. *Ibis*, **90**, 25–45.
Landers, R. Q., Jr. (1962). The influence of chamise, *Adenostoma fasciculatum* in vegetation and soil along chamise–grassland boundaries. Ph.D. Thesis, University of California, Berkeley, Calif.
Larsen, D. W. and K. A. Kershaw (1975). Acclimation in arctic lichens. *Nature*, **254**, 421–423.
Law, R., A. D. Bradshaw, and P. D. Putwain (1977). Life history variation in *Poa annua*. *Evolution*, **31**, 233–246.
Law, R. (1978). The cost of reproduction in annual meadow grass. *Amer. Natur.*, **112** (in press).
Leigh, E. G. (1975). Structure and climate in tropical rain forest. *Ann. Rev. Ecol. Syst.*, **6**, 67–86.
Leith, H. (1960). Patterns of change within grassland communities. In J. L. Harper (Ed.), *The Biology of Weeds*, British Ecological Society Symposium No. 1, Blackwell, Oxford.
Levin, D. A. (1971). Plant phenolics: an ecological perspective. *Amer. Natur.*, **105**, 157–181.
Levin, D. A. (1975). Pest pressure and recombination systems in plants. *Amer. Natur.*, **109**, 437–451.
Levitt, J. (1975). *Responses of plants to environmental stresses*, Academic Press, New York.
Lewis, I. F. (1914). The seasonal life-cycle of some red algae at Woods Hole. *Plant World*, **17**, 31–35.
Lewontin, R. C. (1965). Selection for colonizing ability. In H. G. Baker and G. L. Stebbins (Eds.), *The Genetics of Colonizing Species*, Academic Press, New York and London, pp. 77–91.
Lewontin, R. C. (1971). The effect of genetic linkage on the mean fitness of a population. *Proc. Nat. Acad. Sci.*, **68**, 984–986.
Liddle, M. J. (1975). A selective review of the ecological effects of human trampling on natural ecosystems. *Biol. Conserv.*, **7**, 17–36.
Liddle, M. J. and P. Greig-Smith (1976). A survey of tracks and paths in a sand dune ecosystem: II. Vegetation. *J. appl. Ecol.*, **12**, 909–930.
Lindeman, R. L. (1942). The trophic-dynamic aspect of ecology. *Ecology*, **23**, 399–418.
Livingstone, R. B. and M. L. Allessio (1968). Buried viable seed in successional field and forest stands, Harvard forest, Massachusetts. *Bull. Torrey Bot. Club*, **95**, 58–69.

Lloyd, P. S. (1968). The ecological significance of fire in limestone grassland communities of the Derbyshire Dales. *J. Ecol.*, **56**, 811–826.

Loach, K. (1967). Shade tolerance in tree seedlings: I. Leaf photosynthesis and respiration in plants raised under artificial shade. *New Phytol.*, **66**, 607–621.

Loach, K. (1970). Shade tolerance in tress seedlings: II. Growth analysis of plants raised under artificial shade. *New Phytol.*, **69**, 273–286.

Longman, K. A. and J. Janik (1974). *Tropical Forest and its Environment*, Longman, London.

Loomis, W. D. (1967). Biosynthesis and metabolism of monoterpenes. In J. B. Pridham (Ed.), *Terpenoids in Plants*, Academic Press, London, pp. 59–82.

Loveless, A. R. (1961). A nutritional interpretation of sclerophylly based on differences in chemical composition of sclerophyllous and mesophytic leaves. *Ann. Bot., N.S.*, **25**, 168–176.

Lutz, H. J. (1928). Trends and silvicultural significance of upland forest successions in southern New England. *Yale Univ. School of Forestry Bull.*, **22**, 1–68.

MacArthur, R. H. (1955). Fluctuation of animal populations and a measure of community stability. *Ecology*, **36**, 533–536.

MacArthur, R. H. and E. D. Wilson (1967). *The theory of island biogeography*, Princeton University Press, Princeton, N.J.

MacFarlane, C. and H. P. Bell (1933). Observations of the seasonal changes in the marine algae in the vicinity of Halifax, with particular reference to winter conditions. *Proc. N. S. Inst. Sci.*, **18**, 134–176.

MacFarlane, J. D. and K. A. Kershaw (1977). Physiological–environmental interactions in lichens: IV. Seasonal changes in the nitrogenase activity of *Peltigera canina* (L.) Willd. var. *praetextata* (Floerke in Somm.) Hue, and *P. canina* (L.) Willd. var. *rufescens* (Weiss) Mudd. *New Phytol.*, **79**, 403–408.

Madge, D. S. (1965). Leaf fall and litter disappearance in a tropical forest. *Pedobiologia*, **5**, 272–288.

Mahmoud, A. (1973). A laboratory approach to ecological studies of the grasses *Arrhenatherum elatius* (L.) Beauv. ex J. and C. Presl, *Agrostis tenuis* Sibth. and *Festuca ovina* L. Ph.D. Thesis, University of Sheffield.

Mahmoud, A. and J. P. Grime (1974). A comparison of negative relative growth rates in shaded seedlings. *New Phytol.*, **73**, 1215–1219.

Mahmoud, A. and J. P. Grime (1976). An analysis of competitive ability in three perennial grasses. *New Phytol.*, **77**, 431–435.

Mahmoud, A., J. P. Grime, and S. B. Furness (1975). Polymorphism in *Arrhenatherum elatius* (L.) Beauv. ex J. and C. Presl. *New Phytol.*, **75**, 269–276.

Main, A. R., M. J. Littlejohn, A. K. Lee (1959). Ecology of Australian frogs. In A. Keast, R. L. Crocker, and C. S. Christian (Eds.), *Biogeography and ecology in Australia*, Junk, The Hague, pp. 396–411.

Major, J. and W. T. Pyott (1966). Buried viable seeds in two California bunch grass sites and their bearing on the definition of a flora. *Vegetatio*, **13**, 253–282.

Margalef, R. (1963a). On certain unifying principles in ecology. *Amer. Natur.*, **97**, 357–374.

Margalef, R. (1963b). Successions of populations. *Adv. Frontiers of Plant Sci., Inst. Adv. Sci. Cult., New Delhi, India*, **2**, 137–188.

Margalef, R. (1968). *Perspectives in ecological theory*, Chicago Press, Chicago.

Marks, P. L. (1974). The role of pin cherry (*Prunus pensylvanica* L.) in the maintenance of stability in northern hardwood ecosystems. *Ecol. Monogr.*, **44**, 73–88.

Marshall, R. (1927). The growth of hemlock before and after release from suppression. *Harvest For. Bull.*, **11**, 66pp.

Mason, G. (1976). An investigation of the effect of temperature upon the germination and growth of native species, using temperature-gradient techniques. Ph.D. Thesis. University of Sheffield.

Mason, H. L. (1946). The edaphic factor in narrow endemism: I. The nature of environmental influences. *Madrono*, **8**, 209–226.
McClure, F. A. (1966). Flowering, fruiting and animals in the canopy of a tropical rain forest. *Malay Forest.*, **29**, 182–203.
McLachlan, K. D. (1976). Comparative phosphorus responses in plants to a range of available phosphorus situations. *Aust. J. Agric. Res.*, **27**, 323–341.
McMillan, C. (1956). The edaphic restriction of *Cupressus* and *Pinus* in the Coast Ranges of central California. *Ecol. Monogr.*, **26**, 177–212.
McNaughton, S. J. (1975). r- and K-selection in *Typha. Amer. Natur.*, **109**, 251–261.
McPherson, J. K. and C. H. Muller (1969). Allelopathic effects of *Adenostoma fasciculatum*, 'Chamise', in the California Chaparral. *Ecol. Monogr.*, **39**, 177–198.
McRill, M. (1974). The ingestion of weed seed by earthworms. *Proc. 12th Br. Weed Control Conf.*, **2**, 519–525.
McRill, M. and G. R. Sagar (1973). Earthworms and seeds. *Nature*, **243**, 482.
Medica, P. A., F. B. Turner, and D. D. Smith (1973). Effects of radiation on a fenced population of horned lizards *Phrynosoma platyrhinos* in Southern Nevada. *J. Herpetology*, **7**, 79–85.
Mellinger, M. V. and S. J. McNaughton (1975). Structure and function of successional vascular plant communities in Central New York. *Ecol. Monogr.*, **45**, 161–182.
Menhinick, E. F. (1967). Structure, stability and energy flow in plants and arthropods in a *Sericea lesedeza* stand. *Ecol. Monogr.*, **37**, 255–272.
Miller, R.S. (1967). Pattern and process in competition. *Advance. Ecol. Res.*, **4**, 1–74.
Miller, R. S. (1968). Conditions of competition between redwings and yellow-headed blackbirds. *J. Anim. Ecol.*, **37**, 43–62.
Milne, A. (1961). Definition of competition among animals. In F. L. Milnthorpe (Ed.), *Mechanisms in Biological Competition*, Cambridge University Press, London, pp. 40–61.
Milton, W. E. J. (1939). The occurrence of buried viable seeds in soils at different elevations and in a salt marsh. *J. Ecol.*, **27**, 149–159.
Milton, W. E. J. (1940). The effect of manuring, grazing and cutting on the yield, botanical and chemical composition of natural hill pastures. *J Ecol.*, **28**, 328–356.
Milton, W. E. J. (1943). The yields of ribwort plantain (ribgrass) when sown in pure plots and with grass and clover species. *Welsh J. Agric.*, **27**, 109–116.
Miyata, I. and T. Hosakawa (1961). Seasonal variations of the photosynthetic efficiency and chlorophyll content of epiphytic mosses. *Ecology*, **42**, 766–775.
Monk, C. D. (1966). An ecological significance of evergreenness. *Ecology*, **47**, 504–505.
Monk, C. D. (1971). Leaf decomposition and loss of ^{45}Ca from deciduous and evergreen trees. *Amer. Midl. Natur.*, **86**, 379–385.
Monsi, M. and T. Saeki (1953). Über den lichfaktor in den Pflanzengesell-schaften und seine Bedentung für die Stoffproduktion. *Jap. J. Bot.*, **14**, 22–52.
Montgomery, G. G. and M. E. Sunquist (1974). Impact of sloths on neotropical energy flow and nutrient cycling. In E. Medina and F. B. Golley (Eds.), *Trends in Tropical Ecology: Ecological Studies IV*, Springer, New York.
Mooney, H. A. (1972). The carbon balance of plants. *Ann. Rev. Ecol. Syst.*, **3**, 315–346.
Mooney, H. A. and W. D. Billings (1961). The physiological ecology of arctic and alpine populations of *Oxyria digyna. Ecol. Monogr.*, **31**, 1–29.
Mooney, H. A. and A. T. Harrison (1970). The influence of conditioning temperature on subsequent temperature-related photosynthetic capacity in higher plants. In C. T. de Wit (Ed.), *Prediction and Measurement of Photosynthetic Productivity*, Centre for Agricultural Publishing and Documentation, Wageningen, The Netherlands, pp. 411–417.
Mooney, H. A. and F. Shropshire (1967). Population variability in temperature related photosynthetic acclimation. *Oecol. Plant.*, **2**, 1–13.
Mooney, H. A. and M. West (1964). Photosynthetic acclimation of plants of diverse origin. *Am. J. Bot.*, **51**, 825–827.
Moore, R. J. (1978). Is *Acanthaster planci* an r-strategist? *Nature*, **271**, 56–57.

Morhardt, S. S. and D. M. Gates (1974). Energy-exchange analysis of the belding ground squirrel and its habitat. *Ecol. Monogr.*, **44**, 17–44.

Morse, D. H. (1971). The insectivorous birds as an adaptive strategy. *Ann. Rev. Ecol. Syst.*, **2**, 177–200.

Morse, D. H. (1974). Niche breadth as a function of social dominance. *Amer. Natur.*, **108**, 818–830.

Mortimer, A. M. (1974). Studies of germination and establishment of selected species with special reference to the fate of seeds. Ph.D. Thesis, University College of North Wales.

Mott, J. J. (1972). Germination studies on some annual species from an arid region of Western Australia. *J. Ecol.*, **60**, 293–304.

Muller, C. H. (1940). Plant succession in the *Larrea-Flourensia* climax. *Ecology*, **21**, 206–212.

Muller, C. H. and C. H. Chou (1972). Phyto-toxins: an ecological phase of phytochemistry. In J. B. Harborne (Ed.), *Phytochemical Ecology*, Academic Press, London.

Muller, W. H. and C. H. Muller (1956). Association patterns involving desert plants that contain toxic products. *Amer. J. Bot.*, **43**, 354–361.

Müller, G. and E. Foerster (1974). Entwicklung von Weideansaaten im Überflutungsbereich des Rheines bei Kleve. *Z. Acker und Pflanzenban*, **140**, 161–174.

Munz, P. A. (1959). *A California flora*, University of California Press, Berkeley, California.

Muscatine, L. and E. Cernichiari (1969). Assimilation of photosynthetic products of zooxanthellae by a reef coral. *Biol. Bull.*, **137**, 506–523.

Myerscough, P. J. and F. H. Whitehead (1967). Comparative biology of *Tussilago farfara* L., *Chamaenerion angustifolium* (L) Scop., *Epilobium montanum* L., and *Epilobium adenocaulon* Hausskn: II. Growth and ecology. *New Phytol.*, **66**, 785–823.

Nagel, J. L. (1950). Changement d'essences. *J. Forest Suisse* (*Schwei. Z. Fortswissenschaften*), **101**, 95–104.

Naveh, Z. (1961). Toxic effects of *Adenostoma fasciculatum* (Chamise) in the Californian chaparral. *Proc. 4th Congr. Sci. Soc.* (*Rehovot, Israel*), 1 page.

New, J. K. (1958). A population study of *Spergula arvensis*. *Ann. Bot.*, **22**, 457–477.

Newman, E. I. (1963). Factors controlling the germination date of winter annuals. *J. Ecol.*, **51**, 625–638.

Newman, E. I. (1973). Competition and diversity in herbaceous vegetation. *Nature*, **244**, 310.

Newman, E. I. and A. D. Rovira (1975). Allelopathy among some British grassland species. *J. Ecol.*, **63**, 727–737.

Newton, I. (1964). Bud-eating by Bullfinches in relation to the natural food supply. *J. appl. Ecol.*, **1**, 265–279.

Nichols, J. D., W. Conley, B. Batt, and A. R. Tipton (1976). Temporally dynamic reproductive strategies and the concept of r- and K-selection. *Amer. Natur.*, **110**, 995–1005.

Nicholson, I. A., I. S. Paterson, and A. Currie (1970). A study of vegetational dynamics: selection by sheep and cattle in *Nardus* pasture. In A. Watson (Ed.), *Animal Populations in Relation to their Food Resources*, Blackwell, London, pp. 73–98.

Nieto, J. H., M. H. Brondo, and J. T. Gonzalez (1968). Critical periods of the crop growth cycle for competition from weeds. *Pesticide Articles and News Summaries C*, **14**, 159–166.

Norberg, R. A. (1977). An ecological theory on foraging time and energetics and choice of optimal food searching method. *J. Anim. Ecol.*, **46**, 511–530.

Noy-Meir, I. (1973). Desert ecosystems: environment and producers. *A. Rev. Ecol. Syst.*, **4**, 25–51.

Nye, P. H. (1961). Organic matter and nutrient cycles under moist tropical forest. *Plant and Soil*, **13**, 333–346.

Odum, E. P. (1963). *Ecology*, Holt, Rinehart, and Winston, New York.

Odum, E. P. (1969). The strategy of ecosystem development. *Science*, **164**, 262–270.

Odum, E. P. (1971). *Fundamentals of Ecology*, 3rd ed., Saunders, Philadelphia.

Odum, H. T. and E. P. Odum (1955). Trophic structure and productivity of a windward coral reef community on Eniwetok Atoll. *Ecol. Monogr.*, **25**, 291–320.

Odum, H. T. and R. C. Pinkerton (1955). Time's speed regulator: the optimum efficiency for maximum power output in physical and biological systems. *Amer. Sci.*, **43**, 331–343.

Oechel, W. C. and N. J. Collins (1973). Seasonal patterns of CO_2 exchange in bryophytes at Barrow, Alaska. In L. C. Bliss and F. E. Wielogolaski (Eds.), *Primary Production and Production Processes, tundra biome, proceedings of the conference, Dublin, Ireland*, Swedish IBP Committee, Wenner-Gren Center, Stockholm, Sweden, pp. 197–203.

Ogden, J. (1974). The reproductive strategy of higher plants: II. The reproductive strategy of *Tussilago farfara* L. *J. Ecol.*, **62**, 291–324.

Olmsted, N. W. and J. D. Curtis (1947). Seeds of the forest floor. *Ecology*, **28**, 49–52.

Olsen, J. S. (1963). Energy storage and the balance of producers and decomposers in ecological systems. *Ecology*, **44**, 322–331.

Oosting, H. J. and P. J. Kramer (1946). Water and light in relation to pine reproduction. *Ecology*, **28**, 47–53.

Orians, G. H. and G. Collier (1963). Competition and blackbird social systems. *Evolution*, **17**, 449–459.

Overland, L. (1966). The role of allelopathic substances in the 'smother crop' barley. *Amer. J. Bot.*, **53**, 423–432.

Paine, R. T. (1969). The *Pisaster–Tegula* interaction: Prey patches, predator food preference and intertidal community structure. *Ecology*, **50**, 950–961.

Paine, R. T. (1974). Intertidal community structure: Experimental studies on the relationship between a dominant competitor and its principal predator. *Oecologia* (*Berl.*), **15(2)**, 93–120.

Park, T. (1954). Experimental studies of interspecific competition: II. Temperature, humidity and competition in two species of *Tribolium*. *Physiol. Zoöl.*, **27**, 177–238.

Parsons, R. F. (1968a). The significance of growth-rate comparisons for plant ecology. *Amer. Natur.*, **102**, 595–597.

Parsons, R. F. (1968b). Ecological aspects of the growth and mineral nutrition of three mallee species of Eucalyptus. *Oecol. Plant.*, **3**, 121–136.

Pemadesa, M. A. and P. H. Lovell (1974). Some factors affecting the distribution of some annuals in the dune system at Aberffraw, Anglesey. *J. Ecol.*, **62**, 403–416.

Perring, F. H. (1959). Topographical gradients of chalk grassland. *J. Ecol.*, **47**, 447–481.

Perring, F. H. (1968). *Critical Supplement to the Atlas of the British Flora*, Nelson, Edinburgh.

Perring, F. H. and S. M. Walters (1962). *Atlas of the British Flora*, Nelson, Edinburgh.

Perry, T. O. (1971). Winter-season photosynthesis and respiration by twigs and seedlings of deciduous and evergreen trees. *For. Sci.*, **17**, 41–43.

Pessin, L. J. (1922). Epiphyllous plants of certain regions in Jamaica. *Torrey Bot. Club. Bul.*, **49**, 1–14.

Peterken, G. F. and P. S. Lloyd (1967). Biological Flora of the British Isles: *Ilex aquifolium*. *J. Ecol.*, **51**, 841–858.

Pianka, E. R. (1966). Convexity, desert lizards and spatial heterogeneity. *Ecology*, **47**, 1055–1059.

Pianka, E. R. (1970). On r- and K-selection. *Amer. Natur.*, **104**, 592–597.

Pickering, S. U. (1919). The action of one crop on another. *J. Roy. Hort. Soc.*, **43**, 372–380.

Pigott, C. D. (1955). Biological Flora of the British Isles — *Thymus* L. *J. Ecol.*, **43**, 365–387.

Pigott, C. D. (1968). Biological Flora of the British Isles: *Cirsium acaulon*. *J. Ecol.*, **56**, 597–612.

Pigott, C. D. and K. Taylor (1964). The distribution of some woodland herbs in relation to the supply of nitrogen and phosphorus in the soil. *J. Ecol.*, **52**, 175–185.

Platt, W. J. (1975). The colonization and formation of equilibrium plant species associations in badger disturbances in a tall-grass prairie. *Ecol. Monogr.*, **45**, 285–305.

Polunin, N. (1948). Botany of the Canadian Eastern Arctic, Part III. Vegetation and ecology. *Natl. Mus. Canad. Bull.*, **104**, 304pp.

Prečsenyi, I. (1969). Analysis of the primary production (phytobiomass) in an *Artemisio-Festucetum pseudovinae. Acta Bot. Acad. Hungary*, **15**, 309–325.

Rabotnov, T. A. (1969). On coenopopulations of perennial herbaceous plants in natural coenoses. *Vegetatio*, **19**, 87–95.

Ratcliffe, D. (1961). Adaptation to habitat in a group of annual plants. *J. Ecol.*, **49**, 187–203.

Raunkiaer, C. (1934). *The life forms of plants and statistical plant geography*; being the collected papers of C. Raunkiaer, translated into English by H. G. Carter, A. G. Tansley, and Miss Fansboll, Clarendon, Oxford.

Ray, P. M. and W. E. Alexander (1966). Photoperiodic adaptation to latitude in *Xanthium strumerium. Am. J. Bot.*, **53**, 806–816.

Raynal, D. J. and F. S. Bazzaz (1975). The contrasting life-cycle strategies of three summer annuals found in abandoned fields in Illinois. *J. Ecol.*, **63**, 587–596.

Redmann, R. E. (1975). Production ecology of grassland plant communities in western and north Dakota. *Ecol. Monogr.*, **45**, 83–106.

Rejmánek, M. (1976). Centres of species diversity and centres of species diversification. *Evol. Biol.*, **9**, 393–408.

Reynolds, H. G. (1958). The ecology of the Merriam kangaroo rat (*Dipodomys merriami* Mearns) on the grazing lands of southern Arizona. *Ecol. Monogr.*, **28**, 111–127.

Reynoldson, T. B. (1961). Environment and reproduction in freshwater triclads. *Nature*, **189**, 329–330.

Reynoldson, T. B. (1968). Shrinkage thresholds in freshwater triclads. *Ecology*, **49**, 584–586.

Rhoades, D. F. (1976). The anti-herbivore defences of Larrea. In T. J. Mabry, J. H. Hunziker, and D. R. DiFeo (Eds.), *The Biology and Chemistry of the Creosote Bush, A Desert Shrub*, Dowden, Hutchinson, and Ross, Stroudsburg, Pennsylvania.

Rhoades, D. F. and R. G. Cates (1976). Toward a general theory of plant anti-herbivore chemistry. In J. Wallace (Ed.), *Recent Advances in Phytochemistry, Vol. 10: Biochemical Interactions between Plants and Insects*, Plenum, New York.

Rice, E. L. (1972). Allelopathic effects of *Andropogon virginicus* and its persistence in old fields. *Amer. J. Bot.*, **59**, 752–755.

Rice, E. L. (1974). *Allelopathy*, Academic Press, New York.

Richards, P. W. (1952). *The tropical rainforest*, Cambridge University Press, London.

Ricklefs, R. E. (1977). On the evolution of reproductive strategies in birds: reproductive effort. *Amer. Natur.*, **111**, 453–478.

Ridley, H. N. (1930). *The dispersal of plants throughout the world*, Reeve, Ashford.

Roberts, H. A. (1970). Viable weed seeds in cultivated soils. *Rep. natn. Veg. Res. Stn.* (1970), 25–38.

Roberts, H. A., W. Bond, and R. T. Hewson (1976). Weed competition in drilled summer cabbage. *Ann. appl. Biol.*, **84**, 91–95.

Roberts, H. A. and F. G. Stokes (1966). Studies on the weeds of vegetable crops: VI. Seed populations of soil under commercial cropping. *J. appl. Ecol.*, **3**, 181–190.

Robinson, R. K. (1971). Importance of soil toxicity in relation to the stability of plant communities. In E. Duffey and A. S. Watt (Eds.), *The Scientific Management of Animal and Plant Communities for Conservation*, Blackwell, Oxford, pp. 105–113.

Robinson, T. (1974). Metabolism and function of alkaloids in plants. *Science*, **184**, 430–435.

Rorison, I. H. (1960). Some experimental aspects of the calcicole–calcifuge problem: I.

The effects of competition and mineral nutrition upon seedling growth in the field. *J. Ecol.*, **48**, 585–599.

Rorison, I. H. (1968). The response to phosphorus of some ecologically distinct plant species: I. Growth rates and phosphorus absorption. *New Phytol.*, **67**, 913–923.

Rose, F. (1974). *The epiphytes of Oak*. In M. G. Morris and F. H. Perring (Eds.), *The British oak, its history and natural history*, Classey, Faringdon, pp. 250–273.

Ross, B. A., J. R. Bray, and W. H. Marshall (1970). Effects of long-term deer exclusion on a *Pinus resinosa* forest in north-central Minnesota. *Ecology*, **51**, 1088–1093.

Roughton, R. D. (1972). Shrub age structures on a mule deer range in Colorado. *Ecology*, **53**, 615–625.

Sagar, G. R. and J. L. Harper (1961). Controlled interference with natural populations of *Plantago lanceolata, P. major* and *P. media. Weed Res.*, **1**, 163–176.

Sale, P. F. (1977), Maintenance of high diversity in coral reef fish communities, *Amer. Natur.*, **11**, 337–359.

Salisbury, E. J. (1942). *The reproductive capacity of plants*, Bell, London.

Sarukhan, J. (1974). Studies in plant demography: *Ranunculus repens* L. *R. bulbosus* L. and *R. acris* L: II. Reproductive strategies and seed population dynamics. *J. Ecol.*, **62**, 151–177.

Scaife, M. A. (1976). The use of a simple dynamic model to interpret the phosphate response of plants grown in solution culture. *Ann. Bot.*, **40**, 1217–1229.

Schaeffer, K. and R. Moreau (1958). L'alternance des essences. *Soc. Forest. France-Compté Bull.*, **29**, 1–12, 76–84, 277–298.

Schulz, J. P. (1960). Ecological studies on rainforest in northern Surinam. *Verh. K. Ned. Akad. Wet.*, **53**, 1–367.

Scurfield, G. (1953). Ecological observations in southern Pennine woodlands, *J. Ecol.*, **41**, 1–12.

Sears, J. R. and R. T. Wilce, (1975). Sublittoral, benthic marine algae of southern Cape Cod and adjacent islands: seasonal periodicity, associations, diversity and floristic composition. *Ecol. Monogr.*, **45(4)**, 337–365.

Sernander, R. (1936). Granskär och Fiby urskog. En studie över stormluckornas och marbuskarnas betydelse; den svenska granskogens regeneration (Engl. summary). *Acta Phytogeogr. Suec.*, **8**, 1–232.

Shreve, F. (1942). The desert vegetation of North America, *Bot. Rev.*, **8**, 195–246.

Shure, D. J. (1971). Insecticide effects on early succession in an old-field ecosystem. *Ecology*, **52**, 271–279.

Siccama, T. G., F. H. Bormann, and G. E. Likens (1970). The Hubbard Brook ecosystem study: productivity, nutrients and phytosociology of the herbaceous layer. *Ecol. Monogr.*, **40**, 389–402.

Siegler, D. and P. W. Price (1976). Secondary compounds in plants: primary functions. *Amer. Natur.*, **110,** 101–105.

Singh, J. S. (1968). Net aboveground community productivity in the grasslands at Varanasi. In Symp. on recent advances in tropical ecology: II. In R. Misra and B. Gopel (Eds), *Proc. International Soc. Tropical Ecology*, India, pp. 631–654.

Singh, J. S. and R. Misra (1969). Diversity, dominance, stability and net production in the grasslands at Varanasi, India. *Can. J. Bot.*, **47**, 425–427.

Slatyer, R. O. (1967). *Plant–Water relationships*, Academic Press, London.

Smith, C. C. (1970). The coevolution of pine squirrels (*Tamiasciurus*) and conifers, *Ecol. Monogr.*, **40**, 349–371.

Smith, C. J., J. Elston, and A. H. Bunting (1971). The effects of cutting and fertilizer treatments on the yield and botanical composition of chalk turf. *J. Br. Grassld Soc.*, **26**, 213–223.

Smith, R. H. (1966). Resin quality as a factor in the resistance of pines to bark beetles. In H. D. Gorhold, R. E. McDermott, E. J. Schreiner, and J. A. Winieski (Eds.), *Breeding Pest-Resistant Trees*, Pergamon, New York, pp. 189–196.

Snaydon, R. W. (1962). Microdistribution of *Trifolium repens* L. and its relation to soil factors. *J. Ecol.*, **50**, 133–143.

Southwood, T. R. E. (1976). Bionomic strategies and population parameters. In R. M. May (Ed.), *Theoretical Ecology, Principles and Applications*, Blackwell, Oxford, pp. 26–48.

Southwood, T. R. E. (1977a). Habitat, the templet for ecological strategies? *J. Anim. Ecol.*, **46**, 337–365.

Southwood, T. R. E. (1977b). The relevance of population dynamic theory to pest status. In J. M. Cherrett and G. R. Sagar (Eds), *Origins of Pest, Parasite, Disease and Weed Problems,* Blackwell, Oxford.

Southwood, T. R. E., R. M. May, M. P. Hassell, and G. R. Conway (1974). Ecological strategies and population parameters. *Amer. Natur.*, **108**, 791–804.

Staniforth, R. J. and P. B. Cavers (1977). The importance of cottontail rabbits in the dispersal of *Polygonum* spp. *J. appl. Ecol.*, **14**, 261–267.

Stearns, S. C. (1976). Life-history tactics; a review of the ideas. *Quart. Rev. Biol.*, **51**, 3–47.

Stebbins, G. L. (1952). Aridity as a stimulus to plant evolution. *Amer. Natur.*, **86**, 33–48.

Stebbins, G. L. (1957). Self-fertilization and population variability in the higher plants. *Amer. Natur.*, **91**, 337–354.

Stebbins, G. L. (1971). Adaptive radiation of reproductive characteristics in angiosperms: II. Seeds and seedlings. *Ann. Rev. Ecol. Syst.*, **2**, 237–260.

Stebbins, G. L. (1972). Ecological distribution of centers of major adaptive radiation in angiosperms. In D. H. Valentine (Ed.), *Taxonomy, Phytogeography and Evolution*, Academic Press, London, pp. 7–34.

Stebbins, G. L. and J. Major (1965). Endemism and speciation in the Californian flora. *Ecol. Monogr.*, **35**, 1–36.

Strain, B. R. and V. C. Chase (1966). Effect of past and prevailing temperatures on the carbon dioxide exchange capacities of some woody desert perennials. *Ecology*, **47**, 1043–1045.

Strong, D. R. (1977) Epiphyte loads, tree falls, and perennial forest disruption: a mechanism for maintaining higher tree species richness in the tropics without animals. *J. Biogeog.*, **4**, 215–218.

Swingland, I. R. (1977). Reproductive effort and life-history strategy of the Aldabran giant tortoise. *Nature*, **269**, 402–404.

Sydes, C. (1979). Investigations into the effects of tree litter on herbaceous vegetation. Ph. D. Thesis, University of Sheffield.

Sydes, C. and J. P. Grime (1979). Effects of tree litter on herbaceous vegetation. *J. Ecol.*, **66** (in press).

Tallis, J. H. (1958). Studies in the biology and ecology of *Rhacomitrium lanuginosum* Brid: I. Distribution and ecology, *J. Ecol.*, **46**, 271–288.

Tallis, J. H. (1959). Studies in the biology and ecology of *Rhacomitrium lanuginosum* Brid: II. Growth, reproduction and physiology. *J. Ecol.*, **47**, 325–350.

Tallis, J. H. (1964). Growth studies on *Rhacomitrium lanuginosum. Bryologist*, **67**, 417–422.

Tamm, C. O. (1956). Further observations on the survival and flowering of some perennial herbs: I., *Oikos*, **7**, 273–292.

Tamm, C. O. (1972). Further observations on the survival and flowering of some perennial herbs: II and III. *Oikos*, **23**, 23–28 and 159–166.

Tansley, A. G. (1939). *The British Islands and their Vegetation*, Cambridge University Press, London.

Taylor, R. J. and R. W. Pearcy (1976). Seasonal patterns in the CO_2 exchange characteristics of understory plants from a deciduous forest. *Can. J. Bot.*, **54**, 1094–1103.

Taylorson, R. B. and H. A. Borthwick (1969). Light filtration by foliar canopies; significance for light-controlled weed seed germination. *Weed Sci.*, **17**, 48–51.

Temple, S. (1977). Plant-animal mutualism; coevolution with Dodo leads to near extinction of plant. *Science*, **197**, 885–886.

Tevis, L. (1958). A population of desert ephemerals germinated by less than one inch of rain. *Ecology*, **39**, 688–695.
Thomas, A. S. (1960). Changes in vegetation since the advent of myxomatosis. *J. Ecol.*, **48**, 287-306.
Thomas, M. (1949). Physiological studies in acid metabolism in green plants: I. CO_2 fixation and CO_2 liberation in crassulacean acid metabolism. *New Phytol.*, **48**, 390–420.
Thomas, W. A. and D. F. Grigal (1976). Phosphorus conservation by evergreenness of mountain laurel *Oikos*, **27**, 19–26.
Thompson, K. (1977). An ecological investigation of germination responses to diurnal fluctuations in temperature. Ph.D. Thesis, University of Sheffield.
Thompson, K. and J. P. Grime (1978). Seasonal variation in herbaceous seed banks. *J. Ecol.*, **66** (in press).
Thompson, K., J. P. Grime, and G. Mason (1977). Seed germination in response to diurnal fluctuations of temperature. *Nature*, **67**, 147–149.
Thurston, J. M. (1969). The effects of liming and fertilizers on the botanical composition of permanent grassland, and on the yield of hay. In I. H. Rorison (Ed.), *Ecological Aspects of the Mineral Nutrition of Plants*, Blackwell, Oxford, pp. 3–10.
Tinkle, D. W. (1969). The concept of reproductive effort and its relation to the evolution of life histories of lizards. *Amer. Natur.*, **103**, 501–515.
Tinnin, R. O. and C. H. Muller (1972). The allelopathic influence of *Avena fatua*: the allelopathic mechanism. *Bull. Torrey Bot. Club*, **98**, 287–292.
Tribe, D. E. (1950). The behaviour of the grazing animal – a critical review of present knowledge. *J. Br. Grassld Soc.*, **5**, 200–214.
Turner, F. B., G. A. Hoddenbach, P. A. Medica, and J. R. Lannoms (1970). The demography of the lizard, *Uta stansburiana* Baird and Girard, in southern Nevada. *J. Anim. Ecol.*, **39**, 505–519.
Turner, J. R. (1967). Why does the genotype not congeal? *Evolution*, **21**, 645–656.
Tyler, G. (1971). Studies in the ecology of Baltic seashore meadows: IV. Distribution and turnover of organic matter and minerals in a shore meadow ecosystem. *Oikos*, **22**, 265–291.
Vaartaja, O. (1952). Forest humus quality and light conditions as factors influencing damping-off. *Phytopathology*, **42**, 501–506.
Vaartaja, O. and H. W. Cran (1956). Damping-off pathogens of conifers and of caragana in Saskatchewan. *Phytopathology*, **46**, 391–397.
van Andel, J. and F. Vera (1977). Reproductive allocation in *Senecio sylvaticus* and *Chamaenerion angustifolium* in relation to mineral nutrition. *J. Ecol.*, **65**, 747–758.
van der Maarel, E. (1971). Plant species diversity in relation to management. In E. Duffey and A. S. Watt (Eds), *The Scientific Management of Animal and Plant Communities for Conservation*, Blackwell, Oxford, pp. 45–64.
van der Pijl, L. (1972). *Principles of dispersal in higher plants*, Springer-Verlag, Berlin.
van der Valk. A. G. and C. B. Davis (1976). The seed banks of prairie glacial marshes. *Can J. Bot.*, **54**, 1832–1838.
van der Wall, S. B. and R. P. Balda (1977). Co-adaptations of the Clark's Nutcracker and the pinon pine for efficient seed harvest and dispersal. *Ecol. Monogr.*, **47**, 89–111.
van Dobben, W. H. (1967). Physiology of growth in two *Senecio* species in relation to their ecological position. *Jaarb. I. B. S.*, 75–83.
van Steenis, C. G. G. J. (1958). Rejuvenation as a factor for judging the status of vegetation types: the biological nomad theory. In *Study of Tropical Vegetation*, Proceedings of the Kandy Symposium, UNESCO, pp. 212–215.
van Steenis, C. G. G. J. (1972). *The Mountain Flora of Java*, E. J. Brill, Leiden.
Viereck, L. A. (1966). Plant succession and soil development on gravel outwash of the Muldrow Glacier, Alaska. *Ecol. Monogr.*, **36**, 181–199.
Vogl, R. J. (1973). Ecology of Knobcone pine in the Santa Ana Mountains, California. *Ecol. Monogr.*, **43**, 125–143.

Voight, J. W. (1959). Ecology of a southern Illinois bluegrass–broomsedge pasture, *J. Range Manage.*, **12**, 175–179.

Walker, R. B. (1954). Factors affecting plant growth on serpentine soils. *Ecology*, **35**, 259–266.

Walter, H. (1973). *Vegetation of the Earth in relation to climate and the ecophysiological conditions*, English Universities Press, London.

Wardle, P. (1959). The regeneration of *Fraxinus excelsior* in woods with a field layer of *Mercurialis perennis*. *J. Ecol.*, **47**, 483–497.

Watt, A. S. (1919). On the causes of failure of the natural regeneration in British oak woods. *J. Ecol.*, **7**, 173–203.

Watt, A. S. (1947). Pattern and process in the plant community. *J. Ecol.*, **35**, 1–22.

Watt, A. S. (1955). Bracken versus heather; a study in plant sociology. *J. Ecol.*, **43**, 490–506.

Watt, A. S. (1957). The effect of excluding rabbits from Grassland B (Mesobrometum) in Breckland. *J. Ecol.*, **45**, 861–878.

Watt, A. S. (1960). Population changes in acidophilous grass-heath in Breckland, 1936–57. *J. Ecol.*, **48**, 605–629.

Watt, T. A. (1976). The emergence, growth, flowering and seed production of *Holcus lanatus* L. sown monthly in the field, *Proc. 1976 Br. Crop. Prot. Conf.—Weeds*, 567–574.

Weaver, J. E. and F. W. Albertson (1956). *Grasslands of the Great Plains*, Johnsen, Lincoln, Nebr.

Webb, L. J., J. G. Tracey, and K. P. Haydock (1967). A factor toxic to seedlings of the same species associated with living roots of the non-gregarious subtropical rainforest tree *Grevillea robusta*. *J. appl. Ecol.*, **4**, 13–25.

Webb, L. J., J. G. Tracey, and W. T. Williams (1972). Regeneration and pattern in the subtropical rain forest. *J. Ecol.*, **60**, 675–695.

Welbank, P. J. (1963). Toxin production during decay of *Agropyron repens* (Couch grass) and other species, *Weed Res.*, **3**, 205–214.

Wells, G. J. (1974). The autecology of *Poa annua* L. in perennial ryegrass pastures. Ph.D. Thesis, University of Reading.

Wells, T. C. E. (1967). Changes in a population of *Spiranthes spiralis* (L) Chevall. at Knocking Hoe National Nature Reserve, Bedfordshire, 1962–65. *J. Ecol.*, **55**, 83–99.

Went, F. W. (1948). Ecology of desert plants: I. Observations on germination in the Joshua Tree National Monument, California. *Ecology*, **29**, 242–253.

Went, F. W. (1949). Ecology of desert plants: II. The effect of rain and temperature on germination and growth. *Ecology*, **30**, 1–13.

Went, F. W. (1955). The ecology of desert plants. *Scient. Am.*, **192**, 68–75.

Werner, P. A. (1975). Predictions of fate from rosette size in teasel (*Dipsacus fullonum* L.). *Oecologia (Berl.)*, **20**, 197–201.

Wesson, G. and P. F. Wareing (1969a). The role of light in the germination of naturally-occurring populations of buried weed seeds. *J. Exp. Bot.*, **20**, 401–413.

Wesson, G. and P. F. Wareing (1969b). The induction of light sensitivity in weed seeds by burial. *J. Exp. Bot.*, **20**, 414–425.

Westhoff, V. (1967). The ecological impact of pedestrian, equestrian and vehicular traffic on vegetation. *P-v. Un. int. Conserv. Nat.*, **10**, 218–223.

Westman, W. E. (1975). Edaphic climax pattern of the pygmy forest region of California. *Ecol. Monogr.*, **45**, 109–135.

Whelan, B. R. and D. G. Edwards (1975). Uptake of potassium by *Setaria anceps* and *Macroptilium atropurpureum* from the same standard solution culture. *Aust. J. Agric. Res.*, **26**, 819–829.

Whitmore, T. C. (1975). *Tropical rain forests of the Far East*, Oxford University Press, London.

Whittaker, E. and C. H. Gimingham (1962). The effects of fire on regeneration of *Calluna vulgaris* (L) Hull from seed, *J. Ecol.*, **50**, 815–822.

Whittaker, R. H. (1966). Forest dimensions and production in the Great Smoky Mountains. *Ecology*, **47**, 103–121.

Whittaker, R. H. (1975). *Communities and Ecosystems*, 2nd Ed. Macmillan, New York.
Whittaker, R. H. and P. P. Feeny (1971). Allelochemics: chemical interactions between species, *Science*, **171**, 757–770.
Wilbur, H. M., D. W. Tinkle, and J. P. Collins (1974). Environmental certainty, trophic level, and resource availability in life history evolution. *Amer. Natur.*, **108**, 805–817.
Wilde, S. A. and D. P. White (1939). Damping-off as a factor in the natural distribution of pine species. *Phytopathology*, **29**, 367–369.
Williams, G. C. (1966). Natural selection, the costs of reproduction, and a refinement of Lack's principle. *Amer. Natur.*, **100**, 687–692.
Williams, G. C. (1975). *Sex and Evolution*. Princeton University Press, Princeton.
Williamson, P. (1976). Above-ground primary production of chalk grassland allowing for leaf death. *J. Ecol.*, **64**, 1059–1075.
Willis, A. J. (1963). Braunton Burrows: the effects on the vegetation of the addition of mineral nutrients to the dune soils. *J. Ecol.*, **51**, 353–374.
Wilson, E. O. (1971). Competitive and aggressive behavior. In W. Dillon and J. F. Eisenberg (Eds). *Man and Beast*, Smithsonian Institution. Washington, D.C., pp. 183–217.
Wilson, R. E. and E. L. Rice (1968). Allelopathy as expressed by *Helianthus annuus* and its role in old-field succession. *Bull. Torrey Bot. Club*, **95**, 432–448.
Woods, D. B. and N. C. Turner (1971). Stomatal response to changing light by four tree species of varying shade tolerance. *New Phytol.*, **70**, 77–84.
Woodwell, G. M. and A. L. Rebuck, (1967). Effects of chronic gamma radiation on the structure and diversity of an oak-pine forest. *Ecol. Mongr.*, **37**, 53–69.
Wourms, J. P. (1972). The developmental biology of annual fishes: III. Pre-embryonic and embryonic diapause of variable duration in the eggs of annual fishes. *J. Exp. Zool.*, **182**, 389–414.
Yemm, E. W. and A. J. Willis (1962). The effects of maleic hydrazide and 2, 4-dichlorophenoxyacetic acid on roadside vegetation. *Weed Res.*, **2**, 24–40.
Zobel, D. B. (1969). Factors affecting the distribution of *Pinus pungens*, an Appalachian Endemic. *Ecol. Mongr.*, **39**, 303–333.

Index of plant and animal names

Page numbers in *italics* refer to references in tables or figures

Subject index